ADVANCES IN
Immunology

EDITED BY

F. J. DIXON, JR.
Division of Experimental Pathology
Scripps Clinic and Research Foundation
La Jolla, California

HENRY G. KUNKEL
The Rockefeller University
New York, New York

VOLUME 8

1968

ACADEMIC PRESS  New York and London

ACADEMIC PRESS INC.
111 Fifth Avenue, New York, New York 10003

United Kingdom Edition published by
ACADEMIC PRESS INC. (LONDON) LTD.
Berkeley Square House, London W.1

LIBRARY OF CONGRESS CATALOG CARD NUMBER: 61–17057

PRINTED IN THE UNITED STATES OF AMERICA

LIST OF CONTRIBUTORS

Numbers in parentheses indicate the pages on which the authors' contributions begin.

D. W. DRESSER, *National Institute for Medical Research, London, England* (129)

DAVID A. LEVY, *Department of Radiological Sciences, The Johns Hopkins University School of Hygiene and Public Health, Baltimore, Maryland* (183)

LAWRENCE M. LICHTENSTEIN, *Department of Medicine, The Johns Hopkins Hospital, Baltimore, Maryland* (183)

N. A. MITCHISON, *National Institute for Medical Research, London, England* (129)

GÖRAN MÖLLER, *Department of Bacteriology, Karolinska Institutet Medical School, Stockholm, Sweden* (81)

HANS J. MÜLLER-EBERHARD, *Department of Experimental Pathology, Scripps Clinic and Research Foundation, La Jolla, California* (1)

ABRAHAM G. OSLER, *Department of Microbiology, The Johns Hopkins University School of Medicine, Baltimore, Maryland* (183)

JONATHAN W. UHR, *Irvington House Institute and Department of Medicine, New York University School of Medicine, New York, New York* (81)

PREFACE

The appearance of Volume 8 so soon after the publication of Volume 7 reflects the increasing number of subjects in which the recent accumulation of information makes a critical review essential for continued rapid progress. It also indicates a commendable willingness on the part of authorities in the field to take the time to sum up the state of affairs in their particular areas of interest.

This volume covers subjects ranging from mechanisms involved in the regulation of antibody formation and in the induction of immunological paralysis to the basic chemistry of some of the humoral participants in immunological injury and, finally, to an *in vitro* analysis of allergy in man. Each chapter is written by an author, or authors, well recognized for outstanding research in the field.

In the first chapter, Dr. Hans Müller-Eberhard presents an authoritative summary of our current knowledge of the chemistry and reaction mechanisms of complement. With definition of eleven distinct serum proteins in the complement system, the anatomy of this system seems reasonably clear. Isolation of many of these components now makes possible the analysis of the chemistry and dynamics of the complement reactions themselves and the understanding of the protein–protein interactions and enzyme activations involved. The application of this chemical information to the study of the biological consequences of complement action is already underway. The effects of complement, primarily on cell surface membranes, may eventuate in a spectrum of changes ranging from cell lysis to directed migration, histamine release, and susceptibility to phagocytosis, all of which can now be partially described in molecular terms.

The second chapter, by Jonathan Uhr and Göran Möller, deals with the regulatory effect of antibody on antibody formation. As we learn more about the process of antibody synthesis it becomes possible to analyze those factors which may influence it. Certainly antibody, the end product of the process, is among the most potent and specific of inhibitors of antibody snythesis. That this inhibition results from the interaction of antibody with antigen neutralizing the immunogenicity of the latter seems likely, and the evidence for this is critically presented. The potential usefulness of this mechanism is suggested by its effectiveness in the enhancement of tissue grafts, its use in the therapeutic prevention of anti-D antibody responses in mothers of Rh-incompatible fetuses, and its possible role in the induction of some types of immunological tolerance.

Drs. D. W. Dresser and N. A. Mitchison present in the third chapter the current status of our understanding of immunological paralysis. It appears clear that the paralysis is the result of a central failure of responsiveness brought about by exposure to antigen which prevents potentially competent cells from initiating a specific antibody response. In cellular terms, paralysis seems to result from an adverse effect of antigen directly on lymphocytes, and the recovery from paralysis appears to involve the recruitment of new competent cells, not a recovery of responsiveness by once paralyzed cells. The quantitative factors involved in this antigenic exposure are now being determined, and these data not only shed light on the process of immunological paralysis but also on those conditions essential to the initiation of an antibody response.

The last chapter by Abraham Osler, Lawrence Lichtenstein, and David Levy describes the development and application of an *in vitro* system in the study of human reaginic allergy. The system utilizes human leukocytes as indicators which release histamine when reaginic antibody fixed to them interacts with its specific allergen. In this system, blocking or nonreaginic antibody can also be assayed by its interference with histamine liberation. Study of allergic patients by this means indicates that severity of their disease is directly related to their cell-bound reagin and demonstrates the protective effects of blocking antibody; it also allows an immunological characterization of the course of the patient.

We thank the authors for their care and thought in preparing this group of scholarly chapters which are certain to stimulate further thought and investigation. To the publisher we express our appreciation for the understanding and cooperation essential to the success of this serial publication.

FRANK J. DIXON
HENRY G. KUNKEL

December, 1967

CONTENTS

Chemistry and Reaction Mechanisms of Complement

HANS J. MÜLLER-EBERHARD

Regulatory Effect of Antibody on the Immune Response

JONATHAN W. UHR AND GÖRAN MÖLLER

The Mechanism of Immunological Paralysis

D. W. Dresser and N. A. Mitchison

In Vitro Studies of Human Reaginic Allergy

Abraham G. Osler, Lawrence M. Lichtenstein,
and David A. Levy

Contents of Previous Volumes

ADVANCES IN

Immunology

VOLUME 8

Chemistry and Reaction Mechanisms of Complement[1]

HANS J. MÜLLER-EBERHARD

Department of Experimental Pathology, Scripps Clinic and Research Foundation, La Jolla, California

[1] This is publication No. 216 from the Department of Experimental Pathology, Scripps Clinic and Research Foundation, La Jolla, California. This work was supported by United States Public Health Service Grant 7007 and by American Heart Association Grant 65-G-166.

1

I. Introduction

During the past decade, developments in immunology have led to an increased awareness of the biological importance of antibody. Concomitantly, it has become apparent that antibody per se is biologically largely ineffective unless aided by effector systems. Complement constitutes the principal, immunologically relevant effector system that is present in blood serum. It consists of nine components or eleven distinct serum proteins. Membranes are the primary target of complement. They may be irreversibly damaged, sustaining distinct ultrastructural lesions, by direct attack which requires participation of all nine complement components. Or they may be otherwise affected by interaction with only certain components or split products thereof. Depending on the cell type involved, such noncytolytic reactions of complement may result in histamine release, directed cellular migration, or increased susceptibility to phagocytosis.

In recent years complement research has experienced an explosive development. The anatomy of this multicomponent system is practically concluded. The chemistry of the components and the dynamics of the complement reaction are now under investigation. At the same time, efforts are being directed toward delineation of the biology of complement. The concept emerging from current work visualizes protein–protein interactions and the formation of complex enzymes as the chemical basis for complement function. Accordingly, throughout the subsequent discussion emphasis will be placed on those observations that relate to the molecular concept, i.e., the chemical and enzymatic nature of complement components and their functionally relevant interactions. The concluding part of the discussion will be devoted to observations which relate to the biological relevance of complement. For the final aim of complement research must be the integration of the newly acquired chemical information into the overall concept of immunology.

II. Nomenclature

The complement nomenclature used in this publication conforms with the recommendations of the active members of the Complement Workshop held in La Jolla, 1966. Individual complement components are designated by numbers. Roman letters will be used to refer to subunits and fragments of components and to indicate their activation and inactivation. The conventional designation of the three components acting in the initial steps of immune cytolysis have been retained. The six later acting components will be designated according to their sequence of action. Thus, the entire system is represented sequentially in the following manner: C′1, C′4, C′2, C′3, C′5, C′6, C′7, C′8, C′9. The three subunits of C′1 are denoted C′1q, C′1r, C′1s.

Intermediate products occurring during the reaction of complement (C′) with erythrocyte(E)–antibody(A) complexes are described systematically according to their composition or according to the reaction steps they have completed: EAC′1a, EAC′1a,4, EAC′1a,4,2a, etc. The small letter "a" indicates the fact that C′1 and C′2 undergo activation in becoming hemolytically effective. A site (S) at the surface of an EAC′1a cell that has reacted with A and C′1 is called SAC′1a. A region of the cell membrane that, through the action of complement, has sustained functional and structural impairment is called S*; E* is a cell with at least one S*.

Products of complement components which arise from their interaction in cellfree solution or which accumulate during immune cytolysis in the fluid phase may be described as follows: C′2i and C′4i are the hemolytically inactive forms of C′2 and C′4, respectively, which are produced by the action of C′1 esterase (see below); C′3i is the hemolytically inactive form of C′3 produced by action of the C′4,2a complex (see below); C′4,2 is a reversible protein–protein complex of hemolytically active C′2 and C′4; and C′4i,2 represents such a complex of C′2 and C′4i. Both of these latter complexes are devoid of enzymatic activity. By contrast, (C′4i,2)a represents a stable protein–protein complex which is derived from C′2 and C′4 by the action of C′1 esterase in the presence of Mg^{++} and which possesses the capacity to convert enzymatically C′3 to C′3i. The complex is incapable of becoming bound to receptor groups owing to the effect of C′1 esterase on its C′4-derived subunit, but it can induce hemolysis from the fluid phase although with very low efficiency. However, C′4,2a is the symbol denoting the same active complex in its receptor-bound form. It exhibits high cytolytic efficiency if bound to cell surface receptors.

So far, three distinct enzymatic activities have been recognized to be associated with the complement system. C'1 esterase refers to an enzyme activity residing in the C'1s subunit of C'1. It is capable of hydrolyzing synthetic esters and it has two natural substrates, C'2 and C'4. C'3 convertase is the trivial name of the fluid phase complex (C'4i,2)a, and the receptor-bound complex, C'4,2a. This enzyme cleaves C'3. The term C'3-dependent peptidase refers to an activity which is apparently resident in a C'4,2a,3 site which hydrolyzes dipeptides containing an aromatic amino acid and the natural substrate of which is presumably one or more of the later acting components.

Two alternative nomenclatures (see Table I) are being used for the six late acting components of guinea pig complement. R. A. Nelson (1965)

TABLE I
COMPLEMENT NOMENCLATURE

Source	Components (listed according to sequence of action)								
Recommended by Complement Workshop, La Jolla, 1966	C'1	C'4	C'2	C'3	C'5	C'6	C'7	C'8	C'9
R. A. Nelson (1965)	C'1	C'4	C'2	C'3c	C'3b	C'3e	C'3f	C'3a	C'3d
Klein and Wellensiek (1965a)	C'1	C'4	C'2	C'3a	C'3b		C'3β	C'3c	C'3d

has called these C'3c,b,e,f,a,d, and Klein and Wellensiek (1965a), C'3a,α,b,β,c,d. Nelson's terminology corresponds directly to the numerical symbols: C'3, C'5, C'6, C'7, C'8, C'9.

III. Physical, Chemical, and Immunochemical Properties of Complement Components

A. HIGHLY PURIFIED COMPONENTS

1. Human C'1q

Table II summarizes the properties of both highly and partially purified human complement components. C'1q represents one of the three subunits of the first component of complement (Müller-Eberhard and Kunkel, 1961; Müller-Eberhard, 1961; Taranta et al., 1961; Hinz and Mollner, 1962; Barbaro, 1963; Lepow et al., 1963). It occurs in human serum in a concentration of 100–200 μg./ml. Its sedimentation coefficient, $s_{20,w}^{\circ}$, is 11.1 S, the sedimentation velocity being markedly concentration dependent, suggesting considerable asymmetry of the molecule. Upon filtration through Sephadex G-200 the protein is excluded by

TABLE II
Properties of Human Complement Components

Properties	C'1q	C'1r	C'1s	C'2	C'3	C'4	C'5	C'6	C'7	C'8	C'9
Serum conc. (μg./ml.)	100–200	—	22	10	1200	430	75	—	—	—	1–2
Sedimentation rate	11.1 S	7 S	4 S	5.5 S	9.5 S	10 S	8.7 S	5–6 S	5–6 S	8 S	4.5 S
Approx. mol. wt.	400,000	—	—	115,000	240,000	230,000	—	—	—	79,000	—
Electroph. mobil.	γ_2	β	α_2	β_2	β_1	β_1	β_1	β_2	β_2	γ_1	α
Carbohydrate conc. (%)	17	—	—	—	2.7	14	19	—	—	—	—
Reactive SH	—	—	—	1 or more	2	—	—	—	—	—	—

the gel and is eluted with the external volume; similarly, it does not enter polyacrylamide or starch gels. Its molecular weight is estimated to be 400,000–500,000. On Pevikon block electrophoresis in phosphate buffer, pH 6, the protein migrates toward the cathode as a sharply defined, homogeneous component. Chemical analysis revealed that it contains approximately 17% carbohydrate, i.e., neutral hexose and hexosamine, but no neuraminic acid, and that 17% of its peptide moiety consists of glycine (Calcott and Müller-Eberhard, 1968). Antisera to C'1q have been obtained and their reactivity with C'1q was not influenced by absorption with various immunoglobulins (Morse and Christian, 1964; Calcott and Müller-Eberhard, 1968). Treatment with 0.1 M mercaptoethanol was without effect on the sedimentation behavior; however, mercaptoethanol and urea dissociate the molecule into subunits. Urea starch gel electrophoresis in formate buffer, pH 3, of reduced and alkylated C'1q revealed two protein bands. The major component migrated only a short distance toward the cathode, whereas the minor component moved considerably faster and closely resembled in mobility and distribution light chains derived from γG-globulin. Immunochemical studies of the subunits are required to determine whether there are any antigenic similarities between C'1q and the immunoglobulins. A relation to the immunoglobulins is suggested by the finding of markedly reduced C'1q levels in agammaglobulinemia (Müller-Eberhard and Calcott, 1968). C'1q exhibits an antibody-like specificity for various γG- and γM-globulins with which it enters into reversible interaction (Müller-Eberhard and Calcott, 1966). It is inhibited by polyinosinic acid, presumably by entering into reversible complex formation with this compound (Yachnin et al., 1964). Since guinea pig C'1 was found to be synthesized in the small intestines (Colten et al., 1966), it may be inferred that the C'1q subunit also originates from this organ. This is of interest in view of the fact that the immune system evolved from intestinal tissue (Good et al., 1966). On the other hand Stecher et al. (1967) demonstrated synthesis of C'1q by macrophages.

2. Human C'1s

This subunit of the first component was shown by Haines and Lepow (1964a) to have a sedimentation coefficient of approximately 4 S and to migrate on paper and immunoelectrophoresis as an α_2-globulin. C'1s carries the esterolytic site of C'1, is present in serum as a proenzyme, and is activated on interaction of C'1 with immune complexes (Becker, 1959; Lepow et al., 1963; Haines and Lepow, 1964b). The concentration of C'1s in human serum is approximately 22 μg./ml. Specific antibody has been produced in rabbits and was shown to be inhibitory to the enzyme (Haines and Lepow, 1964c).

3. Human C'2

The concentration of C'2 in human serum is estimated to be of the order of 10 μg. per ml. It has now become possible to isolate this protein in a highly homogeneous form which on polyacrylamide gel electrophoresis yields a single, sharp protein band. C'2 hemolytic activity could be shown to correspond exactly to this protein zone and previous treatment with C'1 esterase, which converts C'2 to C'2i, resulted in a shift of the protein toward the anode (Polley and Müller-Eberhard, 1968). C'2 has a sedimentation rate of 5.5 S, a molecular weight of 115,000, and it is a β-globulin. There is indirect evidence for the existence of at least one reactive sulfhydryl group in the C'2 molecule which is critical for the activity. C'2 can be partially inactivated by treatment with para-hydroxymercuribenzoate (Leon, 1965) or with p-CMB. Prior treatment with iodine, which enhances C'2 hemolytic activity (see below), renders the molecule resistant to the p-CMB effect (Polley, 1966).

4. Human C'3

This complement component is present in human serum in a concentration of approximately 1.2 mg./ml., which is by far the largest amount of any complement component in serum (Lundh, 1964; West et al., 1964; Klemperer et al., 1965; Kohler and Müller-Eberhard, 1967). It is readily demonstrated by immunoelectrophoresis of whole human serum, the corresponding precipitin arc being located in the β-globulin region and partially within the transferrin arc. On the basis of its immunoelectrophoretic behavior, it has previously been called β_{1C}-globulin (Müller-Eberhard et al., 1960). Its sedimentation coefficient, $s^0_{20,w}$, is 9.5 S and the molecular weight is estimated to be approximately 240,000. The amino acid composition is unremarkable. There is 2.7% carbohydrate in the protein, and serine is the only amino terminal, amino acid residue demonstrated so far (Budzko and Müller-Eberhard, 1968). The molecule contains two free sulfhydryl groups which were detected by binding studies with ^{14}C-labeled parachloromercuribenzoate (p-CMB). After reduction with 10^{-3} M dithiothreitol, four additional sulfhydryl groups became detectable by this method (Polley and Müller-Eberhard, 1967a). The protein can be readily labeled with radioactive iodine without loss of hemolytic activity (McConahey and Dixon, 1966; Alper et al., 1966; Müller-Eberhard et al., 1966a). In serum as well as in the purified state it is inactivated by treatment with 0.02 M hydrazine for 1 hour at 37°C.

(Taylor and Leon, 1959; Müller-Eberhard, 1961; Pondman and Peetoom, 1964), 0.5 M potassium thiocyanate at 4°C. for several hours (Dalmasso and Müller-Eberhard, 1966), or dialdehyde dextran (Fjellström and Müller-Eberhard, 1968). At least three distinct, antigenic determinants have been identified in the native C'3 molecule, which have been called A, B, and D (West et al., 1966). Specific antiserum to C'3 may be produced simply by injecting rabbits with zymosan particles that were incubated at 37°C. with fresh serum and then thoroughly washed (Mardiney and Müller-Eberhard, 1964).

Upon treatment with EAC'1a,4,2a or with the free solution complex (C'4i,2)a, the C'3 is converted to C'3i, its hemolytically inactive reaction product (Pondman and Peetoom, 1964; Müller-Eberhard et al., 1966a). The s rate of C'3i is 8–9 S and the converted protein migrates in the electric field as an α_2-globulin. It contains three reactive sulfhydryl groups, as determined by its capacity to bind p-CMB-^{14}C. It lacks one antigenic determinant which is present in the native molecule (B) and it contains another determinant (Dd) apparently lacking in native C'3 (West et al., 1966). At pH 5 or below, C'3i dissociates a small fragment, the molecular weight of which is estimated to be between 8000 and 20,000 (Dias Da Silva and Lepow, 1966a; Müller-Eberhard et al., 1967a). This fragment is endowed with biological activity (see below). The low molecular weight fragment is now called F(a)C'3 and the residual, major portion of the C'3 molecule is denoted F(b)C'3. In whole serum, C'3i undergoes degradation and the resulting products have been called β_{1A} and α_{2D}. The β_{1A} has a sedimentation coefficient of 6.9 S, a molecular weight of approximately 150,000 and it resembles electrophoretically C'3i (Müller-Eberhard et al., 1960); α_{2D} is electrophoretically faster than β_{1A} and its molecular weight is approximately 70,000. The β_{1A} carries the antigenic determinant A and the α_{2D} carries the determinants D and Dd (West et al., 1966).

Enzymatic and chemical degradation studies have yielded the following preliminary results. Native C'3 is exceedingly susceptible to the action of trypsin; treatment with the enzyme (2% w/w) at room temperature leads within 60 seconds to complete inactivation of C'3 and to conversion to a product closely resembling C'3i. At acid pH both the F(a) and F(b) fragments become demonstrable. If trypsin treatment is extended to 30 minutes, the F(b) fragment is split into two major fragments, which are called F(c)C'3 and F(d)C'3. Their molecular weights are 160,000 and 35,000, respectively, and they are antigenically completely distinct. Treatment of native C'3 with either 0.1 M mercaptoethanol or 6 M urea or with both reagents simultaneously results in aggregation of the protein. Dissociation into subunits, however, may be achieved when the protein is first reduced with $10^{-3} M$ dithiothreitol in the presence of $0.05 M$ dodecyl sulfate, alkylated with iodoacetamide and then exposed to either dodecyl sulfate or glycine buffer, pH 11.8 (Bokisch and Müller-Eberhard, 1968).

Following treatment of whole human serum with mercaptoethanol, C'3 could no longer be detected by immunoelectrophoresis and β_{1A}-globulin assumed the electrophoretic mobility of γ-globulin (Mansa, 1964; Carpenter and Gill, 1966). These observations were interpreted to indicate dissociation of C'3 into subunits by reduction alone. In view of the results with isolated C'3 mentioned above, this conclusion must be regarded with caution and as overinterpretation of data derived solely from immunoelectrophoretic experiments.

Thorbecke and her associates (1965) investigated the site of synthesis of C'3 without arriving at any firm conclusion. The data make it unlikely that peripheral blood cells, lymphocytes, plasma cells, or eosinophiles are responsible. However, macrophages might possibly be the source of C'3 (Stecher et al., 1965).

5. Guinea Pig C'3

Recently, Shin and Mayer (1967) reported the isolation of C'3 from guinea pig serum. The final product was homogeneous on hydroxylapatite chromatography, immunoelectrophoresis, and disc electrophoresis. It is a β-globulin with an s rate of 7.5 S and a molecular weight of 180,000.

6. Human C'4

By immunochemical analysis the concentration of this component in normal human serum was found to be approximately 430 μg./ml. (Kohler and Müller-Eberhard, 1967). The molecular weight of C'4 is 230,000, the $s^{\circ}_{20,w}$ is 10 S. On Pevikon block and immunoelectrophoresis, C'4 behaves as a fast β-globulin. The protein contains 14% carbohydrate which consists of hexose, hexosamine, and neuraminic acid. The amino acid composition is unremarkable. Specific antiserum to C'4 has been prepared and has been used to identify the protein in the immunoelectrophoretic pattern of whole human serum; in the immunoelectrophoretic terminology it is designated β_{1E}-globulin (Peetoom and Pondman, 1963; Müller-Eberhard and Biro, 1963). In serum and in purified form, C'4 is readily inactivated by treatment with dilute ammonia, hydrazine, 0.5 M KCNS (Dalmasso and Müller-Eberhard, 1966), or dialdehyde dextran (Fjellström and Müller-Eberhard, 1968). Inactivation is accompanied by changes in electrophoretic mobility. C'4 can be labeled with radioactive iodine without loss of hemolytic activity (Müller-Eberhand and Lepow, 1965).

Treatment of C'4 with C'1 esterase results in conversion to the hemolytically inactive product C'4i (Lepow *et al.*, 1956a). This material is physicochemically heterogeneous, 60–70% of it having an *s* rate of 9.4 S and 30–40% constituting aggregates of 12–14 S. On gel electrophoresis, C'4i separates into two major bands, one faster, the other slower than native C'4. The fast band corresponds to the 9.4 S material, the slow band to the aggregates (Müller-Eberhard and Lepow, 1965). A low molecular weight product has not been found to result from the conversion. Nor could cleavage of C'4 into two antigenically distinct fragments be confirmed which had been proposed by Peetoom *et al.* (1964). The inactivity of the C'4i molecule arises from its inability to combine with a cellular receptor site; the ability to serve as an acceptor for C'2, however, is unimpaired (Müller-Eberhard *et al.*, 1967a).

Chan and Cebra (1966) isolated C'4 from guinea pig serum and found it to have similar properties to human C'4. From immunohistochemical studies they believe the protein to be synthesized by bone marrow and spleen cells.

7. Human C'5

The concentration of C'5 in human serum is approximately 75 μg./ml. (Nilsson and Müller-Eberhard, 1967a; Kohler and Müller-Eberhard, 1967). The protein is a fast β-globulin and its $s^\circ_{20,w}$ is 8.7 S. The *s* rate increases with increasing protein concentration, which is an anomalous situation and indicates the tendency of C'5 molecules to interact with each other. The protein contains 19% carbohydrate, which consists primarily of hexose and hexosamine with very little neuraminic acid. Specific antiserum has been prepared and has been used to localize C'5 in the immunoelectrophoretic pattern of human euglobulin; in immunoelectrophoretic terms it is referred to as β_{1F}-globulin (Nilsson and Müller-Eberhard, 1965). C'5 can be inactivated in serum or in the purified state by treatment with 1 M KCNS (Dalmasso and Müller-Eberhard, 1966), dialdehyde dextran, or acetic anhydride (Fjellström and Müller-Eberhard, 1968). Brief treatment of C'5 with trypsin (2% w/w) leads to conversion of the protein to an electrophoretically faster component F(b) C'5 and to production of a low molecular weight split product, F(a) C'5 which has biological activity (Cochrane and Müller-Eberhard, 1967a).

8. Human C'9

This is a trace protein of human serum, its concentration being about 1 μg./ml. The *s* rate of C'9 is 4.5 S, the molecular weight 79,000.

The component migrates electrophoretically as a fast α_2-globulin. Homogeneity studies in the analytical ultracentrifuge were not possible so far because of lack of sufficient amounts of C′9. However, on polyacrylamide gel electrophoresis, C′9 yields a single, sharp disc, the position of which corresponds directly to the peak of the activity distribution. Treatment of C′9 with trypsin leads to a 50% enhancement of its hemolytic activity. Using the anthrone method, purified C′9 was shown to contain carbohydrate. The carbohydrate moiety appears to be essential to C′9 function, as periodate treatment led to total inactivation (Hadding et al., 1966).

B. PARTIALLY OR FUNCTIONALLY PURIFIED COMPONENTS

1. Human Components

Intact C′1 was shown by Naff et al. (1964) to have a sedimentation coefficient of 18 S. Upon withdrawal of calcium ions, C′1 dissociates into three subunits, two of which, C′1q and C′1s, were obtained in highly purified form (see above), the third, C′1r, has not been purified so far. It has an s rate of 7 S. C′6 and C′7 are physicochemically highly similar components. Both are slowly migrating β-globulins and both have an s rate of 5 to 6 S (Nilsson and Müller-Eberhard, 1965; Nilsson, 1967). C′8 has an electrophoretic mobility which is intermediate to that of β- and γ-globulin. The sedimentation velocity is approximately 8 S. Production of specific antibodies to these partially purified human complement components has not yet been demonstrated with certainty.

2. Guinea Pig Components

Undissociated C′1 of guinea pig serum behaves as a high molecular weight protein upon filtration on Sephadex G-200. It emerges from the column in the exclusion fraction together with the 19 S globulins (Borsos and Rapp, 1965a). In this respect it resembles human C′1 (Borsos and Rapp, 1965a; Laurell and Siboo, 1966), and, although dissociation has not yet been reported, it probably is also composed of three subunits. Specific antisera to C′1 have been obtained as evidenced by the fact that these antisera caused weak agglutination of EAC′1a but not of EA (Opferkuch and Klein, 1964). The weakness of the reaction was attributed to the reversible nature of the C′1–EA interaction. Recently, Colten et al. (1966) tested a variety of guinea pig tissues for their ability to synthesize C′1 in vitro. Positive results were obtained only with tissue of the small intestines. This finding may be significant in view of the known phylogenetic and ontogenetic derivation of the immune system from this organ (Good *et al.*, 1966).

Guinea pig C′2 was reported to have a sedimentation coefficient of 5.5 S, a diffusion coefficient of 3.9×10^{-7} cm.^{2}/second, a molecular weight of 130,000, and a frictional ratio of 1.60 (Stroud *et al.*, 1966). Borsos and Rapp (1965a) reported a molecular weight of 150,000—a value which was derived solely from gel filtration data. Mayer and Miller (1965) succeeded in raising antisera to C′2 in rabbits and demonstrated their capacity to agglutinate EAC′1a,4,2a and to inhibit C′2 hemolytic activity. The hemolytically inactive product, C′2a^d (Mayer, 1965), which is liberated during the decay of EAC′1a,4,2a to EAC′1a,4, was detected by means of such antisera. The *s* rate of C′2a^d was between 4.3 and 4.7 S, the diffusion coefficient 5.2×10^{-7} cm.^{2}/second, and the molecular weight was thus calculated to be 81,000. Being of distinctly smaller molecular weight than native C′2, the C′2a^d is assumed to constitute a fragment of the native molecule that arises upon C′2 fixation, as will be discussed below. Compared to the native molecule, the fragment exhibits increased electrophoretic mobility and decreased molecular asymmetry (frictional ratio, 1.42). Inactivation of C′2 by C′1a in the fluid phase results in a product which has an *s* rate of 4.5 S (Stroud *et al.*, 1966).

Guinea pig C′3, also called[2] C′3c or C′3a, has an *s* rate of 8.2 S as estimated by density gradient ultracentrifugation (Nelson *et al.*, 1966) and migrates upon paper electrophoresis as a β-globulin (Wellensiek, 1965). Antisera to this component have been prepared which were able to agglutinate EAC′1a,4,2a,3 (Wellensiek *et al.*, 1963; Linscott and Cochrane, 1964), to inhibit the hemolytic activity of bound C′3 (Wellensiek *et al.*, 1963), and to detect C′3 upon immunoelectrophoresis of guinea pig serum (Linscott and Cochrane, 1964). The immunoelectrophoretic appearance is similar to that of human C′3, except that C′3i cannot readily be distinguished from native C′3. C′3 activity is relatively heat resistant, but it is sensitive to hydrazine and to the action of whole cobra venom and to incubation of serum with zymosan (Klein, 1965; Klein and Wellensiek, 1965b). C′5 also called[3] C′3b has an *s* rate of 7.6 S (Nelson *et al.*, 1966) and belongs to the β-globulins (Wellensiek, 1965). It is inactivated by heating at 62°C. and pH 8.5 and by treatment of serum with zymosan or whole cobra venom (Klein, 1965; Klein and Wellensiek, 1965b).

---

[2] Nelson's terminology, C′3c; Klein and Wellensiek's terminology, C′3a.
[3] Nelson's terminology, C′3b; Klein and Wellensiek's terminology, C′3b.

Little information is available on the remaining components except the following. The s rates of C'6, C'7, C'8, and C'9 (corresponding[4] to C'3e, C'3f or C'3β, C'3a or C'3a, and C'3d) have been reported to be 5.7, 5.0, 7.8, and 4.5 S, respectively (Nelson *et al.*, 1966). All these components, except C'7, which has not been examined, are known to be β-globulins as determined by paper electrophoresis (Wellensiek, 1965). This includes C'9 which in human serum migrates as a fast α-globulin. C'8 appears to be the most heat labile, it is rapidly inactivated upon heating of serum at 56°C. and pH 6.5 (Klein, 1965; Klein and Wellensiek, 1965b). Guinea pig C'4 is also a β-globulin (Wellensiek, 1965; Chan and Cebra, 1966), its s rate was reported to be 7.7 S (Nelson *et al.*, 1966), and the molecular weight 180,000 (Borsos and Rapp, 1965a). Klein and Burkholder (1960) demonstrated that an antibody to C'4 can be prepared and that it causes agglutination of EAC'1a,4.

Reviewing the data now at hand for the human and the guinea pig complement system, the following tentative conclusions may be drawn regarding the nature of the components. All complement components are of macromolecular size, their molecular weights ranging roughly from 60,000 to 400,000. At neutral pH all components are negatively charged and behave as anions, except C'1q, which is the most basic protein of the system. All anionic components belong to the β_1- and β_2-globulin fraction of serum, except human C'1s and human C'9, which migrate as α-globulins. Although certain differences can be discerned, it appears that the physical properties of guinea pig components, as far as they are known, resemble roughly those of their human counterparts. Chemically, only the complement components of human serum have been investigated and these only to a limited extent. As far as determined, they are proteins containing a carbohydrate moiety, which may be sizable, as in the case of C'4 where it accounts for 14% of the mass of the molecule. At least in one instance (C'9) the carbohydrate moiety seems to be essential for the hemolytic activity of the component. None of the components represents a lipoprotein of a density lower than 1.21 (Dalmasso and Müller-Eberhard, 1966). Some complement proteins are strongly antigenic, and specific antisera to them may readily be raised. At least two components (C'3 and C'4) give rise even to the formation of auto- and isoantibodies which are directed toward antigenic determinants hidden in the native protein (Lachmann and Coombs, 1965; Lachmann, 1966). The concentration of individual components in serum varies greatly, such that C'3, for instance, represents

---

[4] Nelson's terminology—C'3e, C'3f, C'3a, C'3d; Klein and Wellensiek's terminology—C'3β, C'3c, C'3d.

a major serum constituent, whereas certain other components constitute trace proteins which to date have evaded recognition by means other than activity assays.

IV. Methods of Preparation of Complement Components

A detailed technical description of the current version of methods used in the author's laboratory for the isolation of C'1q, C'2, C'3, C'4, and C'5 has been rendered elsewhere (Müller-Eberhard, 1967a). The methodology of the purification of C'6, C'7, C'8, and C'9 was published in full: C'6, C'7 (Nilsson and Müller-Eberhard, 1967b; Nilsson, 1967), C'8 (Manni and Müller-Eberhard, 1968), and C'9 (Hadding and Müller-Eberhard, 1968). A schematic summary of the various methods is shown in Fig. 1.

A. Highly Purified Components

1. Human C'1q

This protein may be isolated from serum by making use of its ability to combine with and to precipitate soluble γ-globulin aggregates in the presence of ethylenediaminetetraacetate (EDTA). It is eluted from the washed precipitate at acid pH and high ionic strength (phosphate buffer, pH 5.3, $T/2 = 0.3$) (Müller-Eberhard and Kunkel, 1961). Successful application of this method hinges on the preparation of soluble γ-globulin aggregates, a circumstance which has limited its usefulness in some laboratories.

An alternative method has, therefore, been elaborated which utilizes chromatography, gel filtration, and electrophoresis (Calcott and Müller-Eberhard, 1968). The first step consists of the chromatographic separation of the euglobulins of serum (precipitate obtained at pH 5.4, $T/2 = 0.02$) on carboxymethyl cellulose (CM-cellulose) equilibrated with phosphate buffer, pH 5, $T/2 = 0.1$. The chromatogram is developed by a sodium chloride gradient procedure. C'1q emerges from the column at the end of the chromatogram within a conductance range of 17 to 35 mmhos./cm. It is contaminated with γM- and γG-globulin, the latter being partially aggregated. In the second step, the high molecular weight proteins are separated from γG-globulin by filtration through Sephadex G-200 in phosphate buffer, pH 5.3, $T/2 = 0.3$; C'1q leaves the column with the exclusion fraction. In the third step, C'1q is separated from γM-globulin and other high molecular weight material by electrophoresis in Pevikon blocks using phosphate buffer, pH 6, $T/2 = 0.1$. C'1q is the only protein present that migrates under these conditions

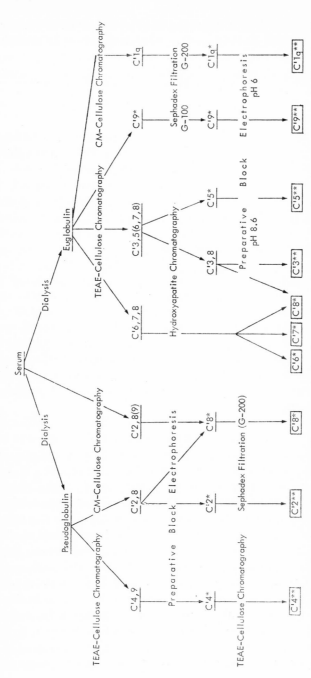

Fig. 1. Schematic presentation of purification methods of complement components from human serum. (* Functionally pure; ** highly purified.)

toward the cathode. The distribution of C'1q hemolytic activity is identical to the distribution of the narrow, cathodally migrating protein peak. As an average, 5–10 mg. of C'1q are obtained from 500 ml. serum. In the analytical ultracentrifuge, a single, sharp component is seen, sedimenting with a velocity of 10 to 11 S.

2. Human C'1s

The method for the purification of this enzyme was developed by Haines and Lepow (1964a). It consists of three chromatographic steps and utilizes the euglobulin fraction as starting material which is precipitated from serum upon dilution with acetate buffer, pH 5.5, $T/2 = 0.02$. Before chromatography the enzyme is activated by exposing the euglobulins in phosphate buffer, pH 7.4, $T/2 = 0.15$, for 15 minutes at 37°C. The material is then applied to a diethylaminoethyl (DEAE)-cellulose column equilibrated with the same phosphate buffer, and the chromatogram is developed with an NaCl gradient. C'1 esterase is eluted at ionic strength 0.37–0.41. The esterase peak fractions of several DEAE chromatograms are pooled and subjected in the second step to chromatography on triethylaminoethyl (TEAE)-cellulose in 0.02 M glycine buffer, pH 9.0, using an NaCl gradient. The active material is then rechromatographed on TEAE-cellulose under essentially identical conditions. Final purification was reported to be 2400-fold with respect to whole serum and the purified esterase had a specific activity of 3800 units/mg. N. The yield was 1.1 mg./liter of serum or 5% of the amount of enzyme available in serum.

3. Human C'2

Purification of this component is a difficult task because of the marked lability of its activity. Two different methods have been elaborated but not yet reported in full (Lepow et al., 1965a; Polley and Müller-Eberhard, 1965a). The procedure used in the author's laboratory is based on the following principles. As C'2 is readily inactivated by C'1 esterase (Lepow et al., 1956a), and since this enzyme belongs to the euglobulins, the latter are removed from serum by precipitation in 0.008 M EDTA solution, pH 5.4. Apparently, some C'1 esterase remains in the pseudoglobulin fraction where it will cause reduction of C'2 activity during the first purification step unless it is inactivated by treatment of the pseudoglobulins with 10^{-3} M diisopropyl fluorophosphate (DFP) for 1 hour at 37°C. DFP is also added to the starting material of each of the later steps. Throughout the procedure, buffers of pH 6 containing 0.002 M Na₃ EDTA are employed, since C'2 was found to be most stable under these conditions. The pseudoglobulin fraction is first separated on CM-cellulose in phosphate buffer, pH 6, $T/2 = 0.06$, using an NaCl gradient elution technique. Fractions containing C′2 activity are rechromatographed under the same conditions and then subjected to electrophoresis on a Pevikon block in phosphate buffer, pH 6, $T/2 = 0.05$. C′2 is recovered from the β-globulin region and is further purified by chromatography on hydroxylapatite employing phosphate buffer, pH 5.8. After adsorption of C′2 to the column in phosphate buffer having a conductance of 3 mmhos./cm., the column is washed with buffer of a conductance of 14 mmhos./cm. and C′2 is then eluted with 16 mmhos./cm. phosphate buffer (pH 5.8). The final degree of purification is approximately 5000-fold with respect to serum and the yield is approximately 2.5 mg./1000 ml. serum. The final product is homogeneous by polyacrylamide electrophoresis (Polley and Müller-Eberhard, 1968).

4. Human C′3 and C′5

Both of these proteins are precipitated from serum with the euglobulin fraction which is used as starting material and which is obtained by dialysis of serum against $0.008\,M$ Na₃EDTA solution of pH 5.4. The euglobulin is separated by TEAE-cellulose chromatography using a salt and pH gradient elution procedure (Müller-Eberhard et al., 1960; Nilsson and Müller-Eberhard, 1965). A fraction is thus obtained containing primarily C′3 and C′5, which is further processed by chromatography on hydroxyapatite equilibrated with phosphate buffer, pH 7.9, and a conductance of 8 mmhos./cm. (Nilsson and Müller-Eberhard, 1965). Stepwise elution is carried out by raising the phosphate concentration while maintaining constant pH. C′5 is eluted at 12 mmhos./cm. and C′3 at 14.5 mmhos./cm., following a wash with buffer of a conductance of 13 mmhos./cm. The hydroxyapatite step is essential as resolution of C′3 and C′5 into two completely separate fractions could not be achieved by any other protein separation procedure currently in use. Hydroxyapatite was made according to Tiselius et al. (1956), commercially available preparations were not of satisfactory quality for this purpose. Final purification of C′3 and C′5 is accomplished by electrophoresis in barbital buffer, pH 8.6, $T/2 = 0.05$, on Pevikon blocks. In this step a small amount of contaminating γ-globulin is removed from C′5 and traces of C′8 are eliminated from C′3. Both C′3 and C′5 are homogeneous upon polyacrylamide electrophoresis, their activity corresponding to the protein band. As an average, 60–80 mg. of C′3 and 5–10 mg. of C′5 are obtained from 1000 ml. of serum.

A simple and rapid method capable of yielding partially purified C′3 was reported by Steinbuch *et al.* (1963). Human plasma is adsorbed with bentonite to remove fibrinogen and lipoproteins. It is then treated with 0.5% Rivanol and the ensuing precipitate containing C'3 is washed, dissolved, and dialyzed against a low ionic strength acetate buffer, pH 5.3. The resulting precipitate consists primarily of C'3. Schultze *et al.* (1962) elaborated a method for the isolation of β_{1A}-globulin, utilizing Rivanol and ammonium sulfate precipitation and preparative zone electrophoresis. The final product was homogeneous by immunochemical, ultracentrifugal, and starch gel electrophoresis criteria.

5. Human C'4

The pseudoglobulin fraction of serum is first separated by chromatography on TEAE-cellulose equilibrated with 0.02 M phosphate buffer, pH 7.3. The column is washed with buffer containing 0.11 M NaCl. When the conductance of the effluent has reached 12 mmhos./cm., elution of C'4 by a NaCl gradient is started. The C'4-containing material is concentrated and further separated by electrophoresis on Pevikon blocks in barbital buffer, pH 8.6, $T/2 = 0.05$. The β-globulin fraction is recovered and subjected again to chromatography on TEAE-cellulose using NaCl gradient elution. The final product is homogeneous on analytical polyacrylamide gel electrophoresis. Usually 10–15 mg. is obtained from 500 ml. of serum (Müller-Eberhard and Biro, 1963).

6. Human C'9

Since the ninth component is a trace protein in human serum, its isolation is cumbersome. The euglobulin fraction of 1000 ml. serum is first separated by chromatography on TEAE-cellulose, the procedure being identical with the first step of the preparation of C'3 and C'5. The C'9 activity is eluted at the end of the chromatogram at a conductance of 25 mmhos. The concentrated material is then passed through a column of Sephadex G-100 equilibrated with Veronal–sodium chloride buffer, pH 7.3, $T/2 = 0.15$. In this step, C'9, which has a relatively low molecular weight, is separated from a variety of serum proteins of 7 to 12 S size. The behavior of C'9 on Sephadex is similar to that of albumin. The active material is then subjected to preparative electrophoresis on a Pevikon block in phosphate buffer, pH 6.0, $T/2 = 0.05$. It migrates more rapidly than the contaminating proteins and is recovered from the α-globulin region in highly purified form. The yield is between 100 and 200 μg. from 1000 ml. of serum. The concentrated component presents itself as one protein band on polyacrylamide gel electrophoresis; C'9 activity corresponds directly to this protein (Hadding and Müller-Eberhard, 1968).

B. Partially Purified Components

1. Human Components

Of the human complement components, only four proteins have not yet been obtained in a high degree of purity. These are C'1r, C'6, C'7, and C'8, all of which can readily be obtained in functionally pure form. C'1r is obtained according to Lepow *et al.* (1963); C'6, C'7 and C'8 may be prepared from the euglobulin fraction following separation on TEAE-cellulose analogous to the procedure used in the first step of

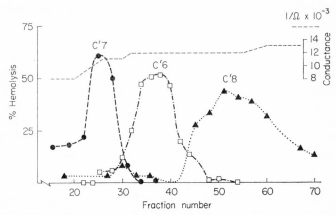

Fig. 2. Curves showing separation of C'6, C'7, and C'8 by means of hydroxy-apatite chromatography. Starting material: a chromatographic fraction obtained by separation of euglobulin on TEAE-cellulose. The material contains C'6, C'7, C'8, and γ-globulin and is devoid of C'3, C'5, and C'9. Starting buffer: phosphate, pH 7.9; conductance, 8 mmhos./cm. Stepwise elution: phosphate, pH 7.9 and conductance of 11, 12, and 13 mmhos./cm., respectively. (Nilsson and Müller-Eberhard, 1967b.)

the preparation of C'3 and C'5 (see above). C'6, C'7, and C'8 leave the column in the early part of the chromatogram before the pH has reached 6.8 and the conductance 6 mmhos./cm. Column fractions of a conductance range of 5.4 to 6 mmhos./cm. are pooled and applied to a column of hydroxyapatite equilibrated with phosphate buffer, pH 7.9, and a conductance of 8 mmhos./cm. By increasing the phosphate concentration, the conductance is raised stepwise to 11, 12, and 13 mmhos./cm. In this fashion the three components may be quantitatively separated, C'7, C'6, and C'8 emerging in this order at conductances of 11, 12, and 13 mmhos./cm., respectively (Nilsson and Müller-Eberhard, 1967b; Nilsson, 1967) (Fig. 2). Examination of the fiftyfold concentrated preparations by immunodiffusion tests and by polyacrylamide gel electrophoresis revealed the presence of at least two serum constituents in each instance.

A method for the isolation of C′8 has been developed recently (Manni and Müller-Eberhard, 1968), however, extensive testing of the final product has not been concluded. Whole human serum is used as starting material and is subjected to precipitation with Rivanol at a final concentration of 0.5% at pH 7.5. Most of the γ-globulins remain in the supernatant under these conditions, whereas C′8 is precipitated together with 70% of the serum proteins. The dissolved precipitate is separated by CM-cellulose chromatography at pH 6 in phosphate buffer using an NaCl gradient. C′8 is eluted at the end of the chromatogram at a conductance of 6 to 12 mmhos./cm. More than 99% of the contaminating protein is eliminated by this step, but the C′8 material contains some C′2 and C′9, in addition to some γ-globulin. The two contaminating activities and most of the γ-globulin are removed by Pevikon block electrophoresis in phosphate buffer, pH 6, $T/2 = 0.05$. The final step consists of filtration through Sephadex G-200 in phosphate buffer, pH 6. On polyacrylamide gel electrophoresis the final product yields one band to which the C′8 activity corresponds and two faster moving proteins which are inactive.

2. Guinea Pig Components

Extending the studies by Linscott and Nishioka (1963), Nelson and his associates elaborated a detailed method for the partial purification of nine complement components from guinea pig serum (Inoue and Nelson, 1965, 1966; Nelson et al., 1966). The purity criteria chosen by Nelson to safeguard preparation of satisfactory reagents are as follows: (a) the reagents should have a low protein content, preferably less than 0.06 (O.D. 280); (b) a high reactivity of the individual component in hemolytic assays, at least 1000 CH_{50} units/ml.; and (c) a reactivity of no more than two other components at dilutions of less than 1:10. No attempt has been made to characterize the final products physicochemically or immunochemically as to their composition.

The first step of the method consists of the precipitation of a low-solubility euglobulin fraction at pH 7.5 and ionic strength 0.04. The precipitate contains C′1 which may be further purified by reprecipitation during dialysis against phosphate buffer, pH 7.5, and 0.005 M. The supernatant which contains all complement components except C′1 is then subjected to chromatography on CM-cellulose equilibrated with 0.06 M sodium chloride, pH 5. The chromatogram is developed with a sodium chloride gradient and the sequence and the ionic strength of elution of the eight components are: C'2, 0.05; C'3f, 0.12; C'4, 0.13; C'3c, 0.135; C'3b, 0.14; C'3e, 0.145; C'3a, 0.18; and C'3d, 0.20. Although a definite sequence of elution is discernible, it is not surprising that the various activities are grossly overlapping and that at this stage the degree of separation is exceedingly limited. Pools from this chromatogram serve as starting materials for eight additional chromatographic steps through which the various reagents are obtained and which satisfy the above cited criteria. Column No. 2 serves the purification of C'2, which is performed on CM-cellulose equilibrated with pH 5.5 and 0.035 M sodium chloride. Column No. 3 serves the purification of C'3f which is done on CM-cellulose equilibrated with 0.055 M sodium chloride, pH 5. After elimination of C'2 by washing the column with 0.065 M sodium chloride, a sodium chloride gradient is applied to elute C'3f. In column No. 4, C'4-containing material is further fractionated on DEAE cellulose which is equilibrated with 0.07 M sodium chloride, pH 7.5 (DEAE). To eliminate C'3f the column is first washed before C'4 is eluted by sodium chloride gradient. Further separation of C'3c is done by DEAE chromatography in column No. 5 which is equilibrated with 0.04 M sodium chloride, pH 7.5. An initial wash step with 0.045 M sodium chloride eliminates C'3e. The C'3c is eluted with 0.075 M sodium chloride, and contaminating C'4 and C'3b are retained by the column under these conditions. Since C'3c at this stage was still substantially contaminated with C'3e and C'3f activity, it was once again chromatographed (column No. 6), this time on CM cellulose, equilibrated with .095 M sodium chloride, pH 5. After washing the column with 0.095 M sodium chloride the final C'3c preparation was obtained by salt gradient elution. In column No. 7, C'3b was subjected to further purification. This column was charged with DEAE-cellulose in 0.07 M sodium chloride at pH 7.5. C'2, C'3c,e and f were removed by washing with starting buffer. C'3b was then eluted by a gradient of increasing sodium chloride concentration and decreasing pH. Under these conditions, some of the contaminating C'4 emerged prior to C'3b. Column No. 8 was used to separate C'3e from contaminating C'3c. Diethylaminoethyl cellulose, 0.04 M sodium chloride, and pH 7.5, were employed to elute C'3e, while C'3c remained adsorbed. To obtain reagents containing C'3a and C'3d, it was necessary to begin with a separate serum sample. After removal from serum of two precipitates formed at 0.04 M sodium chloride and pH 7.5 and 0.06 M sodium chloride, pH 5, the supernatant was subjected to DEAE chromatography at pH 7.5. The column was first washed with 0.145 M sodium chloride, and C'3a and C'3d were then eluted by a salt gradient. A second chromatographic step on DEAE-cellulose was necessary to separate the two activities.

Klein (1965; Klein and Wellensiek, 1965b) reported the chromatographic separation of five of the late acting components in guinea pig serum. Diethylaminoethyl cellulose was used, and the activities were eluted by sodium chloride, ranging from 0.02 to 0.2 M at pH 6.5. The sequence of elution of the five factors was C′3c, β, a, d, b (Klein's nomenclature, see above). By filtration of guinea pig serum through Sephadex G-200, C′3d was found to be eluted with albumin and to be separated from C′3a, b, β, and c which emerged earlier in the fractions containing proteins of intermediate size. C′3d was thus shown to be smaller in molecular size than any of the other late acting components. Klein (1965) also prepared reagents lacking one or more of the five terminal components then recognized. Ammonia- or hydrazine-treated sera lack C′3a but contain C′3b,β,c, and d. Cobra venom treated serum was devoid of C′3a and b, but contained C′3c, β, and d. Heating of serum at 56°C. and pH 7 selectively inactivated C′3c; heating at 57°C. and pH 6.5 eliminated C′3β and c; and heating at 62°C. and pH 8.5 removed C′3b, β, and c. Fischer (1965) obtained partially purified C′9 from pig serum. An ammonium sulfate cut (30–70% saturation) was subjected to gel filtration on Sephadex G-200 and the material of the last eluted peak was further separated by DEAE chromatography. A bimodal distribution of the activity in the chromatogram was observed.

V. Mechanism of Action of Complement in Immune Cytolysis

A. Ultrastructural Manifestations of the Effect of Complement

Goldberg and Green (1959), working with Krebs ascites tumor cells, showed that cells treated with antibody and complement rapidly undergo drastic morphological changes. Within 2 minutes the cells begin to swell and large areas of cytoplasmic matrix become cleared of particulate structures. Mitochondria and the endoplasmic reticulum also undergo swelling and give rise to the formation of large cytoplasmic vesicles. The morphological changes are the consequence of a permeability defect of the outer cell membrane caused by complement. Within 5 minutes after onset of complement action, 90% of the intracellular potassium was found to egress, whereas extracellular sodium and water entered the cytoplasm; the influx of water being caused by high intracellular colloid osmotic pressure. Thus, the initial lesion which permits only small ions to pass is "stretched" sufficiently to allow macromolecules to leave the cell. After 1 to 2 hours, 50% of the intracellular protein and 75% of intracellular ribonucleic acid (RNA) and nucleotides have leaked into the supporting medium (Green *et al.*, 1959a). According to Green and associates (1959b), immune cytolysis represents a form of colloid osmotic lysis.

When Humphrey and Dourmashkin (1965) examined in the electron microscope the membranes of cells that were lysed by antibody and complement, using the negative staining technique, they found distinct ultrastructural lesions of round or slightly ovoid appearance which had a diameter of 80 to 100 Å. The defects, which were filled with negative stain and surrounded by a less electron dense, irregular rim of 20 Å. thickness, were scattered without any apparent order over the membrane surface. Since they looked like holes, they were interpreted to be holes. Lesions could be observed in membranes of sheep erythrocytes which were sensitized with Forssman antibody and treated with guinea pig complement, or which were passively sensitized with *Shigella shigae* "O" somatic antigen and then treated with rabbit anti-*Shigella* antibody and guinea pig complement. They were also seen in the membranes of Krebs ascites tumor cells lysed by rabbit antibody against Krebs cells and guinea pig complement. Borsos *et al.* (1964a) attributed considerable significance to the lesions because they found that the number of lesions per cell corresponded roughly to the predicted number of damaged sites per cell, as estimated on the basis of the one-hit theory of immune hemolysis (Mayer, 1961a). There was little doubt, therefore, in the minds of these workers that the "holes" represent the sites of damage produced by complement.

Extending these observations to human erythrocytes, Rosse *et al.* (1966) found that the mean diameter of the lesion was 103 Å. when human complement was used, and 88 Å. when guinea pig serum was employed. The nature or source of the antibody did not influence the size of the lesion nor did the cell type. The size of the lesion thus appears to be dependent solely upon the type of complement used. They further found that the postulated one hit–one hole hypothesis does not hold true for human complement. Whereas each complete activation of guinea pig complement apparently leads to the production of a single "hole" in the erythrocyte membrane, each complete activation of human complement gives rise to a cluster of "holes." Lysed by an excess of human complement, human erythrocytes showed a maximum of 90,000 lesions per cell. Partial lysis caused by a limited amount of complement was accompanied by the appearance of lesions in the lysed cells, but not in the unlysed cells which were present in the same reaction mixture.

This fact is consistent with the hypothesis that the lesions are indicative of complement-dependent membrane damage. Their chemical and ultra-structural nature remains obscure. Although they have the appearance of holes and present themselves as indentations on the edge of a membrane fragment, it is improbable that they constitute perforations of the membrane. Furthermore, it is not known whether they mark the site of the functional impairment of the membrane and the escape route of intracellular constituents, including macromolecules.

B. Schematic Description of Mechanism

During the past 10 years a large body of information on the mechanism of action of complement has been accumulated. The results obtained by several groups of investigators may be schematically summarized as follows (see Fig. 3): After antibody (A) has combined with

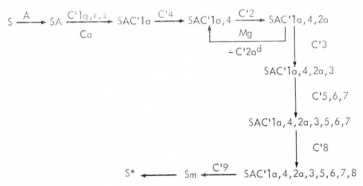

FIG. 3.   Schematic presentation of membrane damage by complement.

an antigenic site (S) on the cell surface, interaction of the first component with bound antibody leads to activation of C'1. Thus is set in motion the chain of events which involves all nine components and results in the production of a damaged site, S*, in the cell membrane. Although there are nine components, there are fewer than nine functional units, probably only five, which may be correlated, with five major phases of the complement reaction. In the first phase C'1 interacts with antibody and C'1 esterase is thereby activated. In the second phase an enzymatically active complex (C'4,2a) is assembled on the cell surface which is derived from the second and fourth components. In the third phase, C'3 becomes attached to the cell surface and this event is accompanied by the appearance of peptidase activity. During the fourth phase, C'5, C'6, and C'7 impart their effect on the cell membrane, rendering it permanently susceptible to the action of C′8 and C′9. In the final phase, C′8 and C′9 activate a membrane "effector," which, in turn, causes the impairment of membrane function which becomes manifest in overt cytolysis.

C. The Role of Antibody

1. Dependence of Cytolytic Capacity of Antibody on Immunoglobulin Type

Antibody per se does not interfere with the structural and functional integrity of cell membranes. The role of antibody in immune cytolysis may be understood in terms of its ability to select specifically the target of complement action. As will be seen below, antibody increases the cytolytic efficiency of complement several hundred-fold. Thus, although it does not appear to be an essential factor in cell damage by complement, it fulfills the function of an adapter permitting the first component of complement to act in close proximity to the cell surface. Not all antibodies are able to bind C′1 and thus to mediate immune cytolysis, and those antibodies that can bind C′1 may vary greatly in their cytolytic capacity. γA-Type antibodies appear to lack the ability to interact with C′1 (Ishizaka et al., 1966a; Borsos and Rapp, 1966), and the hemolytic efficiency of antibodies of the γM type is far greater than that of antibodies belonging to the γG class (Talmage et al., 1956; Stelos and Talmage, 1957). Although three of the four known heavy-chain subgroups of human γG-globulin were shown to fix complement, these subgroups may differ quantitatively in their complement-fixation efficiency. Benacerraf et al. (1963) could separate two different types of guinea pig antibody of the γG-class by electrophoresis. Whereas the slowly migrating antibody was able to interact with complement and thus to mediate cell lysis, the electrophoretically fast traveling variety lacked this capacity.

It is now clear that only one antibody molecule of γM type and two of γG type are required to form a potentially cytolytic site (Borsos and Rapp, 1965b). Originally, a greater number of antibody molecules were thought to be needed, since the rates of hemolysis with γM and with γG antibody were found to be a function of the square and of the fourth power of the antibody concentration, respectively (Weinrach et al., 1958). These findings were interpreted to indicate that sensitization of a site on the cell surface to complement action required two molecules of γM antibody or four molecules of γG antibody. According to Borsos and Rapp (1965c), this interpretation is based on the erroneous assumption that the rate of action of complement is the same for all concentrations of antibody.

Humphrey and Dourmashkin (1965) correlated the number of antibody molecules bound per cell with the number of ultrastructural lesions produced in the presence of an excess of complement. Using radioactively labeled antibody they found that for the production of one ultrastructural lesion two to three molecules of γM antibody were sufficient, whereas 2000–3000 molecules of γG type were required. Since there are 90,000 sites per sheep erythrocyte with which a γM antibody can combine, it is improbable that the two or three molecules would attach to the same site. It was concluded, therefore, that a single γM-antibody molecule which is attached to the cell surface must be sufficient to activate complement and to produce one lesion. Measurements and calculations for γG antibody were compatible with the assumption that two neighboring molecules are required to form one complement reactive site. Recently, Borsos and Rapp (1965b) applied the C′1 fixation and transfer test to the enumeration of antibody molecules required for the formation of a lytic site. They analyzed their data by plotting the logarithm of the number of "effective" C′1a molecules fixed as a function of the logarithm of relative antibody concentration. The slopes of the resulting lines were considered a measure of the number of antibody molecules required to establish one site capable of binding one molecule of C′1a. For γM antibody the slope was 1 and for γG antibody it was 2.1 to 2.6. Thus, utilizing titration data, Borsos and Rapp confirmed Humphrey's results which were based on physical and immunochemical evidence.

Within the γG-globulin class there are presently recognized four distinct subgroups designated: γG₁ (2b, We), γG₂ (2a, Ne), γG₃ (2c, Vi), and γG₄ (2d, Ge) (Grey and Kunkel, 1964; Terry and Fahey, 1964; Ballieux et al., 1964). The first three subgroups were tested for their complement-fixing ability by Ishizaka et al. (1966b) who aggregated respective myeloma proteins with bisdiazotized benzidine and found all three groups reactive. So far γG₄ myelomas have not been tested by this technique. It will be shown below, however, that this subgroup shows an exceedingly low reactivity with C′1q.

2. The Complement-Combining Region in γG-Globulin

Elucidation of the gross structure of γG-globulin has afforded the exploration of the combining region for complement in the molecule. Initially, the complement combining site was localized to the Fc fragment (Taranta and Franklin, 1961; Ishizaka et al., 1962). However, Amiraian and Leikhim (1961) found residual complement-fixing ability in pepsin-treated antibody, i.e., the $F(ab')_2$ piece. Schur and Becker (1963) extended these observations by demonstrating that pepsin-treated antibody in the form of precipitates with antigen were able to fix 40% of guinea pig complement. The remaining 60% could readily be depleted by precipitates formed with 7 S antibody but resisted fixation with a fresh precipitate made with 5 S antibody. This finding suggested that guinea pig complement was heterogeneous with respect to its first component, one portion being capable of reacting with 5 S antibody as well as 7 S antibody, and the other portion being reactive only with 7 S antibody. It also suggests the possible occurrence of two distinct sites for C'1 in the intact γG molecule—one located in the Fc and the other in the Fab portion.

As shown by Wiedermann et al. (1963, 1964), the complement-fixation capacity of γG-globulin may be strikingly reduced by treatment with mercaptoethanol. A quantitative analysis of the phenomenon by Schur and Christian (1964) showed that in the case of rabbit antibody only one interchain disulfide bond, or a total of two disulfide bonds per molecule, need to be cleaved to reduce complement-fixing efficiency to approximately 30%. It thus appears that the integrity of the region of the labile, inter-H chain, disulfide bond is critical for the interaction of γG-globulin with complement. Although papain action is thought to take place near this region, cleavage of antibody with insoluble papain did not impair its ability to react with complement (Cebra, 1963). Lately, Griffin et al. (1967) found that labeling of a few tryptophan residues in antidinitrophenyl antibody with 2-hydroxy-5-nitrobenzyl bromide decreased the complement reactivity of the molecule substantially without affecting the antibody-combining site. The location of these critical residues within the molecule has not yet been explored.

Although the C'1-combining site of γG-globulin has not been clearly delineated, it is apparent that its location has to be sought in the heavy chains and evidently not only in their Fc portion. In addition to the C'1 site, there are also sites for C'3 and C'4, since both components have been shown to be able to combine physically with γ-globulin (Müller-Eberhard, 1962; Peetoom and Pondman, 1963; Leddy et al., 1963, 1965; Willoughby and Mayer, 1965; Müller-Eberhard and Lepow, 1965). According to Chan and Cebra (1966), C'4 attaches preferentially to the Fab portion of antibody, and there is some suggestive evidence that this may also apply to C'3 (Reiss and Plescia, 1963).

3. Effect of Antibody Specificity on Immune Cytolysis

There is evidence that specificity of an antibody may influence its complement-binding efficiency. For instance, although most human antibodies are capable of complement fixation, certain antibodies to blood group antigens, such as anti-Rh antibodies, lack this ability (Adinolfi *et al.*, 1962; Polley and Mollison, 1961; Stratton, 1961; Harboe *et al.*, 1963). It is difficult to imagine that the inability is an inherent feature of the antibody itself. It is more likely to find the explanation in the antigen-dependent spacial distribution of these antibodies on the red cell surface. Studies by several groups of investigators revealed that there are approximately 5000–30,000 D antigen sites per cell, the exact number depending upon the genotype of the individual (Rochna and Hughes-Jones, 1965; Boursnell *et al.*, 1953; Masouredis, 1960). By contrast, there are approximately one million A antigen sites (Kabat, 1956) or 500,000 I antigen sites per human red cell (Rosse *et al.*, 1966). In the case of the sheep erythrocyte, there are at least 600,000 Forssman antigen sites per cell (Humphrey and Dourmashkin, 1965). The small number of Rh antigen sites per cell may cause these sites to be too far apart to permit antibody molecules to be sufficiently close for complement activation. In support of this view, Klun and Muschel (1966) found that noncomplement-fixing cell–antibody systems could be rendered complement fixing when previously lysed cells were employed. Apparently the relative density of antigenic sites on cell stromata is somewhat greater than on the corresponding intact cells.

The specificity of antibodies may also have an impact on the size of the functional lesion caused by complement. In experiments with human erythrocytes, Sears *et al.* (1964) found that the effective radius of the lesion was 32 Å. when rabbit, antihuman, erythrocyte antibody plus human complement were used, and greater than 32.5 Å. when the lesions were produced with rabbit or human antihuman A and human complement. These measurements were made with macromolecules of known effective diffusion radius and were based upon their ability to protect antibody- and complement-treated cells against colloid osmotic lysis. It should be emphasized that the functional lesions studied by these authors may not be identical with the ultrastructural lesions reported by Humphrey and Dourmashkin (1965). The size of the latter were shown to be independent of antibody and cell type.

D. INTERACTION BETWEEN THE FIRST COMPONENT AND ANTIBODY

1. Structure and Activation of C'1

That C'1 is not a single substance became apparent when C'1q, the 11 S component, was detected (Müller-Eberhard and Kunkel, 1961). Later, Lepow *et al.* (1963) pointed out the existence of a third subcomponent, C'1r, and discovered that the three proteins (C'1q, C'1r, C'1s) occur in serum in the form of a calcium-dependent macromolecular complex. The complex can be dissociated into its subunits by withdrawal of calcium ions, and the subunits may then be separated from each other by chromatography on DEAE-cellulose or by sucrose density gradient ultracentrifugation (Naff et al., 1964). In his studies on the early steps of immune hemolysis, Becker (1959) was led to postulate two functional groups in C'1—a combining site and a catalytic site. Today it is clear that the combining site of the C'1 complex resides in C'1q and the catalytic site in C'1s. Originally, C'1s occurs in the complex as a proenzyme which becomes activated when C'1 interacts with antibody. The mechanism of activation has not been elucidated fully. According to Lepow et al. (1965b), activation proceeds in two reaction steps, the first being considered a slow, spontaneous process, and the second a rapid autocatalytic reaction. Conversion of C'1 to C'1a is temperature dependent (Borsos et al., 1964b) and inhibitable by ε-aminocaproic acid (Taylor and Fudenberg, 1964). It can be achieved by treatment of C'1 with either plasmin or trypsin (Ratnoff and Naff, 1967). If activation through antigen–antibody complexes is an enzymatic process, then C'1q or C'1r should possess enzymatic activity similar to plasmin and trypsin. Such enzyme activity has thus far not been demonstrated for either of these two subcomponents. Ca^{2+} appears to be specifically required for activation of C'1 (Levine et al., 1953). In addition, Ca^{2+} functions as a ligand in the C'1 step and in this capacity it may be replaced by a number of other bivalent cations, such as Zn, Sr, Cd, or Mg (Wirtz and Becker, 1961). Both C'1 and C'1a are inhibitable by carrageenin (Davies, 1963; Borsos et al., 1965) and polyinosinic acid (Yachnin and Rosenblum, 1964; Yachnin et al., 1964). Inhibition by both substances was shown to be due to interference with the combining unit, C'1q.

2. C'1 Transfer

Unlike other complement components, bound and activated C'1 has the ability to transfer from one binding site to the other. Evidence for this phenomenon was first provided by Weigle and Maurer (1957a) who observed transfer of hemolytically active guinea pig C'1 from complement-treated immune precipitates to soluble immune complexes. Becker (1956a) showed that C'1 could be detached from EAC'1 or EAC'1,4 by treatment with 0.01 M EDTA. Upon removal of EDTA and reconstitution of an optimal calcium concentration, the dissociated C'1 could be reattached to EAC'4 to form EAC'1,4. Rapp and Borsos (1963), pursuing observations by others on the effect of salt concentration on complement function (Becker and Wirtz, 1959; Levine et al., 1961; Wardlaw and Walker, 1963), then demonstrated that transfer of C'1 occurs freely at physiological ionic strength and that it can be inhibited by lowering the ionic strength to 0.065. On the basis of these observations, they designed a C'1 fixation and transfer test (Borsos and Rapp, 1965c) which is a highly sensitive method for the detection of immune complexes or for antibody bound to cells. Binding of C'1 is first allowed to take place at low ionic strength and after removal of unbound C'1, transfer of bound C'1 is initiated by raising the ionic strength to 0.15. The EAC'4 serves as C'1 acceptor, and the amount of C'1 transferred is estimated according to standard methods.

3. C'1q–γ-Globulin Interaction

Weigle and Maurer (1957b) showed that complement may precipitate soluble antigen–antibody complexes. They postulated that it is multivalent to antibody and, therefore, able to link the complexes together into precipitating aggregates. In a similar fashion, complement was demonstrated to precipitate soluble aggregates of γ-globulin. Although the responsible serum factor was heat labile, it functioned independent of calcium ions and was thus distinct from the first component of complement (Müller-Eberhard and Kunkel, 1961). The active principle was isolated and was later shown to represent the subcomponent C'1q.

In addition to having affinity for aggregates of γ-globulin and immune precipitates, C'1q is able to interact with native monomeric γG-globulin. The interaction results in the formation of a reversible complex which may readily be demonstrated by analytical ultracentrifugation (Müller-Eberhard and Calcott, 1966). The velocity and the size of the complex increase with increasing concentration of γG-globulin. At saturation the s rate of the complex is 15 S and approximately four molecules of γG-globulin are in association with one molecule of C'1q. The valence of C'1q is, therefore, estimated to be 4. The reaction appears to be highly specific: C'1q reacts with γG- and γM-globulin, but it failed to interact with γA-globulin, Bence-Jones protein, and albumin. In the case of γG-globulin, the extent of the reaction was found to vary according to the heavy-chain subgroup of the protein employed (Müller-Eberhard et al., 1967b). Intense reactions were observed with myeloma proteins of the γG_1 and γG_3 subgroups, a modest reaction with γG_2, and a poor or negative reaction with γG_4-myeloma proteins. Thus, it appears that C'1q is able to differentiate between various classes and subgroups of the γ-globulins, its behavior being reminiscent of that of an antibody. By having specificity for antibody protein while constituting a part of complement, C'1q becomes the functional link between antibodies and their serum effector system. The accumulated evidence indicates that C'1q carries the combining site of the first component and that it triggers the mechanism of activation of C'1 esterase upon combination of C'1 with antibody.

4. Function of C'1

The C'1a bound to EA catalyzes the formation of the next functional unit on the cell surface, the C'4,2a complex. This is accomplished in two separate steps, the first leading to the binding of C'4 to the cell surface, and the second resulting in attachment and activation of C'2. That both steps are mediated by C'1a has been clearly demonstrated by Becker: EAC'1a,4,2a is not formed when C'1 esterase on EAC'1a is inactivated by DFP (Becker, 1960). In addition, Stroud et al. (1965) showed that fixation of C'2 is inhibited by p-toluenesulfonyl-L-arginine methyl ester (TAMe) which is hydrolyzed by C'1 esterase. This fact constitutes further evidence for catalysis of C'2 fixation by activated C'1 esterase.

E. INTERACTION BETWEEN C'2 AND C'4 AND FORMATION OF THE C'3-CONVERTING ENZYME

1. Binding of C'4 to the Cell Surface

To understand the mechanism of C'4 binding it was necessary to explore the effect of C'1 esterase on C'4 (Müller-Eberhard and Lepow, 1965). When highly purified C'4 was treated with highly purified C'1 esterase (1 μg. of C'1 esterase per 5000 μg. of C'4) the following observations were made. C'4 was converted to C'4i, which is hemolytically inactive and differs physicochemically from C'4 in that it has a greater electrophoretic mobility and an s rate of 9.2 instead of 10 S. In addition, C'4i contains 20–40% of aggregated protein which sediments with an s rate of 12 to 14 S. Since the protein–protein complexes were a constant finding, they were considered an essential feature of the effect of C'1 esterase on C'4. C'1 esterase apparently confers upon C'4 an unusual reactivity toward a chemical group occurring in protein, including C'4 itself. To test whether similar groups are available on erythrocyte membranes, C'4 was treated with C'1 esterase in the presence of nonsensitized sheep erythrocytes. Although most of the C'4, as expected, was inactivated in the fluid phase, a small proportion became firmly attached to the cells in hemolytically active form. This finding led to the formulation of the following concept regarding the binding of C'4. Native C'4 is converted by C'1 esterase to a state of activation in which the molecule is enabled to react with a suitable acceptor to become bound. If it fails to collide with an acceptor within a very short period of time, the activated molecule undergoes inactivation and accumulates as C'4i in the fluid phase. An activated form of the C'4 molecule, C'4a, was postulated in view of the inability of C'4 (in absence of C'1 esterase) and of C'4i (regardless of C'1 esterase) to become bound to cell surface receptor groups. The exact chemical nature of the labile intermediate form of the molecule is unknown. The efficiency of C'4 binding is more than 200-fold increased when binding is catalyzed by cell-bound C'1a instead of unbound C'1 esterase present in the fluid phase. Nevertheless, experiments with ^{125}I-labeled C'4 have shown that, even under optimal conditions, only 5–15% of the amount offered is specifically bound to EAC'1a and that the remaining portion is converted in the process to C'4i (Müller-Eberhard et al., 1966a). That C'4 enters into a physical union with those cells with which it reacts was established prior to the use of radioactive labels. Conversion of EAC'1a to EAC'1a,4 was found to be accompanied by the appearance of an agglutinogen that reacted with an anticomplement serum (Klein and Burkholder, 1960). Specific antiserum to highly purified C'4 was shown to agglutinate EAC'1a,4, and this agglutination could be inhibited by addition of isolated C'4 (Müller-Eberhard and Biro, 1963; Harboe et al., 1963). Moreover, cell-bound C'4 could be functionally inhibited by anti-C'4 (Müller-Eberhard and Biro, 1963; Opferkuch and Klein, 1964). By employing radioactively labeled C'4 it was possible to determine the physicochemical characteristics of bound C'4. The component could be dissociated from EAC'1a,4 by treatment with $0.0017 M$ cholate–deoxycholate at pH 7.4. It was found to resemble C'4i with respect to electrophoretic mobility and ultracentrifugal properties. These studies indicate that C'4 is bound during immune cytolysis in macromolecular, activated form probably through hydrophobic bonds (Dalmasso and Müller-Eberhard, 1967).

Regarding the site of binding of C'4 on the cell surface, two possibilities offer themselves—antibody molecules that are attached to the membrane or constituents of the membrane itself. The experiments with nonsensitized sheep erythrocytes cited above have demonstrated that C'4 can bind directly to membrane constituents. Furthermore, Harboe (1964) and Boyer (1964) working with human erythrocytes and cold agglutinins showed that C'4 remains associated with the cells when the antibody is eluted at 37°C. On the other hand, Willoughby and

Mayer (1965) were able to isolate from EAC'1a,4 an antisheep erythrocyte antibody that was firmly complexed with hemolytically active C'4.
Furthermore, formation of complexes of human γG-globulin and C'4
could be demonstrated following incubation of these two proteins with
C'1 esterase (Müller-Eberhard and Lepow, 1965). It appears, therefore,
that C'4 may attach itself either to membrane constituents or to the antibody portion of EA and that in both locations it is cytolytically active.

2. Binding and Release of C'2

The reaction of C'2 with EAC'1a,4 leads to physical uptake of the
C'2 molecule and the concomitant accumulation in the fluid phase of a
hemolytically inactive derivative, C'2i. As has been convincingly demonstrated by Becker (1960), formation of EAC'1a,4,2a from EAC'1a,4 is
catalyzed by cell-bound C'1a. This enzymatic concept was further elaborated by Stroud et al. (1965), who found that fixation of C'2 to
SAC'1a,4 is catalyzed by C'1a in a manner which follows the classic
pattern of enzyme kinetics and that TAMe inhibits the formation of
SAC'1a,4,2a by competing with C'2 for the active site on the enzyme.
Thus, the activating enzyme of C'2 that enables the molecule to become
bound to its cell surface acceptor is C'1 esterase, which, as was pointed
out above, functions also as the activating enzyme for C'4.

Physical uptake of C'2 in the course of formation of EAC'1a,4,2a was
demonstrated by Mayer and Miller (1965) in three different ways.
EAC'1a,4,2a could be agglutinated with a rabbit antiserum to C'2; the
activity of bound C'2a could be inhibited by anti-C'2; and the anti-C'2
titer of the antiserum could be reduced by absorption with EAC'1a,4,2a.
The ability of cell-bound C'2a to adsorb anti-C'2 was found to be
limited and this was interpreted to indicate that antigenic groups of the
C'2 molecule had either become inaccessible in the process of binding
or were lost by cleavage of the molecule.

With the aid of anti-C'2 it became possible to detect hemolytically
inactive C'2, since the latter competes with active C'2 for neutralization
of the antibody. The technique was called by Mayer (1965) "immunologic neutralization-competition test." It was found that treatment of
C'2 with C'1 esterase in cellfree solution, which is known to lead to
inactivation of C'2 (Lepow et al., 1956a), did not destroy its antigenic
properties (Stroud et al., 1965). A similar hemolytically inactive, but
antigenically intact product of C'2 was found to accumulate in the fluid
phase during formation of EAC'1a,4,2a. The precise nature of this antigenic material is not known; in fact, it may constitute a mixture of two or
three different products. Assuming that C'2 is cleaved in the process of

fixation, then C'2i may represent the inert fragment of the molecule that remains unfixed. It may also include that proportion of activated fragments that decays before it has a chance to react with an acceptor. Finally, C'2i contains inactive C'2a, that is, that portion of C'2 which is first bound as C'2a and then is released upon decay of EAC'1a,4,2a (see below). It is estimated that the waste of C'2 ranges between 50 and 90% of the molecules present in a given reaction (Sitomer *et al.*, 1966). This figure includes those molecules that became bound to the cell in a position that renders them cytolytically ineffective.

One of the peculiar features of the complement reaction is the lability of the intermediate product, EAC'1a,4,2a (Mayer *et al.*, 1954), which upon decay reverts to the state EAC'1a,4 (Borsos *et al.*, 1961a). The latter intermediate complex is able to react again with a fresh supply of C'2 and to reform EAC'1a,4,2a. This observation suggested that if C'2 was bound during formation of EAC'1a,4,2a, it was released upon its decay. With the aid of the immunological neutralization-competition test, Stroud *et al.* (1966) found that decay of EAC'1a,4,2a is accompanied by release of an inactive C'2 derivative which they named C'2a^d. The rate of release of this material was essentially the same as the rate of disappearance of C'2a sites from the cells. The electrophoretic mobility of C'2a^d was greater than that of C'2 and its molecular weight was definitely smaller than that of the native molecule. Using gel filtration through Sephadex to determine diffusion constants and ultracentrifugation in a sucrose density gradient to estimate sedimentation coefficients, these authors calculated the molecular weights of C'2a^d and C'2 to be 81,000 and 130,000, respectively.

On the basis of these observations, Mayer and his colleagues (Stroud *et al.*, 1965; Sitomer *et al.*, 1966; Stroud *et al.*, 1966) proposed the following concept. When C'2 is acted on by C'1 esterase, a bond is cleaved which can either react with water or with a suitable group on the cell surface. In the former case, C'2 is inactivated by hydrolysis and, in the latter, it becomes bound in active form. According to this view, C'1 esterase acts as a transferase. Taking into account the molecular weight difference between C'2a^d and C'2, the transferase hypothesis was extended into the cleavage hypothesis. It proposes that cleavage of a bond by C'1 esterase leads to fragmentation of the C'2 molecule with subsequent fixation of the active fragment and release of the inactive fragment into the fluid phase. The latter fragment should have a molecular weight of approximately 49,000. As this fragment has not been found as yet, the cleavage hypothesis remains presently unproven.

3. Formation of a Reversible Complex between C'2 and C'4

Since C'2 is bound to EAC'1a,4 but not to EAC'1a, it has been speculated for some time that C'2 reacts with C'4 on the cell surface. Formation of a molecular complex between C'4 and C'2a has, indeed, been demonstrated recently as will be discussed below. It was revealing, therefore, to find that C'2 and C'4 possess an affinity for each other enabling them to enter into reversible interaction in free solution (Müller-Eberhard et al., 1967a). Sucrose density gradient ultracentrifugation disclosed that C'2 sedimented considerably faster than expected on the basis of its known sedimentation rate when present in mixtures with purified C'4 in which the concentration of C'4 was relatively large. Since a similar variation in sedimentation behavior was not observed with purified C'2 alone, the phenomenon was attributed to complex formation with C'4. The reaction was reversible, association being favored by high pH and low ionic strength. Although stable complexes were formed only above pH 8 at ionic strength 0.1, definite interaction occurred also at physiological pH and ionic strength. Bivalent cations did not seem to be essential.

An analogous reaction apparently occurs between C'2 and cell-bound C'4. Sitomer et al. (1966) have pointed out the capacity of C'2 to become reversibly adsorbed on EAC'4. As in free solution the reaction was favored by high pH and low ionic strength. However, in contrast to the free solution experiments, adsorption of C'2 on EAC'4 was strictly dependent on magnesium ions, 1 mM being the optimal concentration. In spite of this difference, it is most probable that C'2 adsorption to cells is based on its reversible interaction with C'4. Sitomer et al. assume that adsorption constitutes a necessary step of C'2a fixation. They further believe that the magnesium requirement in immune hemolysis is explained by its role in the adsorption process. This explanation may be questioned in view of the fact that C'2–C'4 interaction in free solution proceeds in the absence of magnesium ions.

4. The C'1 Esterase-Induced Complex, C'4,2a

A better understanding of the structure, the mechanism of formation, and mode of action of the functional unit comprising C'2 and C'4 emerged from experiments with purified components that were conducted in cellfree solution. The ultracentrifugal observation of C'2–C'4 interaction was an important clue; however, its functional significance for the complement reaction remained to be demonstrated. It had been known for some time that, in the course of the complement reaction, C'3 becomes converted to C'3i (Müller-Eberhard and Nilsson, 1960) and that conversion is caused only by cells in the state EAC'1a,4, 2a or EAC'4,2a (Pondman and Peetoom, 1964; Linscott and Cochrane, 1964; Müller-Eberhard et al., 1966a). Attempts to produce C'3-converting activity in cellfree solution from purified components showed generation of activity to occur only when a mixture of C'2, C'4, and magnesium were incubated with C'1 esterase (Müller-Eberhard et al., 1967a). Deletion of any of these factors resulted in failure to generate activity. Instead of native C'4, its hemolytically inactive derivative, C'4i, could be employed; however, C'2i was unable to substitute for native C'2. This suggested that C'4i serves as acceptor for C'1 esterase-activated C'2 and that, unless activated C'2 becomes bound to C'4 or C'4i, its activity is irreversibly destroyed.

When the C'3-converting principle was examined by sucrose density ultracentrifugation it was found to have an s rate of 11.2 S, whereas its two precursors, C'2 and C'4i, had s rates of 5.5 and 8.9 S, respectively. Upon filtration through Sephadex G-200, C'3-converting activity emerged from the column before C'4i and C'2. Diffusion coefficients of 3.3, 3.5, and 4.3×10^{-7} cm.^{2}/second were found for the converting principle, C'4i and C'2, respectively. From these data the molecular weight of the C'3-converting enzyme was calculated to be 305,000. This was distinctly larger than the molecular weight of either of its two precursors; the molecular weight of C'4i was 229,000 and of C'2, 115,000. There is little doubt, therefore, that the activity resides in a molecular complex which is derived from C'2 and C'4 by the action of C'1 esterase. It is evident, however, that the sum of the molecular weights of the precursors exceeds the molecular weight of the active complex by 39,000. Provided this difference is not due to limited accuracy of the methods used for the molecular weight determinations, it indicates fragmentation of at least one of the two precursors. Since fragmentation of C'4 by C'1 esterase may be excluded (Müller-Eberhard and Lepow, 1965), fragmentation of C'2 has to be considered. If it occurs, the molecular weight data suggest that the C'2 molecule is cleaved into an activated fragment, F(a) C'2 (molecular weight 76,000), which may become incorporated into the complex, and an inactive fragment, F(b)C'2 (molecular weight 39,000), which is incapable of reacting with C'4. This view is in accord with Mayer's C'2 cleavage hypothesis, verification of which, however, requires demonstration of the inactive fragment.

Quantitative analysis of the reaction between C'3 and the (C'4i,2)a complex indicates that the latter is an enzyme. The enzyme hypothesis is further supported by the finding that the (C'4i,2)a complex cleaves the C'3 molecule: a low molecular weight fragment is dissociated from C'3i, but not from native C'3, as the hydrogen ion concentration is raised (Müller-Eberhard et al., 1967a; Lepow, 1967). The trivial name C'3 convertase has been selected to denote the enzyme. The nature of the bond in C'3, which is cleaved by C'3 convertase, has not been assessed. Although it is probably a peptide bond, a thioester bond has been considered as an alternative (Müller-Eberhard et al., 1966b), since one additional sulfhydryl group becomes detectable in the molecule following conversion to C'3i. It is emphasized that magnesium ions are essential for formation of the enzyme, although the two precursors can interact reversibly in the absence of magnesium. Thus, magnesium requirement in this system cannot be explained by assuming that magnesium functions as ligand between C'2 and C'4 prior to C'2 activation. This explanation was used by Mayer for magnesium requirement in immune hemolysis. The free solution experiments indicate that magnesium fulfills an additional and, possibly, more significant function in the formation of C'3 convertase.

The relevance of the C'4,2a complex in immune hemolysis was demonstrated by the formation of EC'4,2a from EC'4 and purified C'2, C'1 esterase, and magnesium. These cells convert and bind C'3 and can be lysed by EDTA–C' (Müller-Eberhard et al., 1967a).

5. Effect of Iodine Treatment of C'2 on Activity and Stability of C'4,2a

The possible presence of sulfhydryl groups in C'2 that are critical for C'2 activity was first pointed out by Leon (1965), who treated partially purified C'2 with para-hydroxymercuribenzoate and observed partial inactivation. A similar effect was observed with p-CMB but not with iodoacetamide. Initially, treatment of C'2 with iodoacetamide had the opposite effect, namely enhancement of C'2 activity (Polley and Müller-Eberhard, 1965b). When freshly recrystallized iodoacetamide was used, neither inactivation nor enhancement were observed. It was soon found that iodine, which is usually a contaminant of iodoacetamide, was solely responsible for enhancement of the activity (Polley, 1966; Polley and Müller-Eberhard, 1966). Treatment of partially purified human C'2 at a protein concentration not exceeding 1 mg./ml. with an equal volume of $10^{-4} M$ I_2 in $5 \times 10^{-3} M$ KI at pH 6 and 0°C. for 5 minutes led to a 5–13-fold increase of C'2 hemolytic activity as measured by effective molecule titration (see below). When iodine-treated C'2 was used for the formation of EAC′1a,4,2a the half-life of this intermediate complex at 32°C. was found to be increased from approximately 13 minutes to more than 150 minutes. A similarly drastic increase in activity and stability was found for the free solution (C′4i,2)a complex if iodine-treated C′2 was used for its preparation (Müller-Eberhard et al., 1967a). In fact, demonstration of this complex after 20 hours of density gradient ultracentrifugation became possible only through exploitation of the iodine effect.

In exploring the nature of the critical modification caused by iodine, it was found that enhancement of C′2 hemolytic activity did not correlate with iodination of the C′2 molecule (Polley and Müller-Eberhard, 1967b). As an alternative explanation of the effect, oxidation of a sulfhydryl group was therefore considered. Oxidation to disulfide seemed unlikely since mild reduction with mercaptoethanol did not reverse the iodine effect, which, however, could be reversed by sodium dithionite. Since iodine-treated C′2 can no longer be inactivated by p-CMB, it is believed that iodine and p-CMB act on the same critical sulfhydryl group, which when oxidized by iodine, becomes inert to p-CMB. On the basis of this hypothesis, it was expected that p-CMB-inactivated C′2 is resistant to enhancement by iodine. Contrary to expectations, iodine was found not only to restore the original activity of p-CMB-treated C′2, but also to cause the usual enhancement. The simplest explanation for this phenomenon is that oxidation with iodine leads to release of p-CMB. Although this hypothesis has not yet been tested for C′2, the validity of this principle was explored by using C′3, which can bind 2 molecules of p-CMB per molecule of protein. With ^{14}C-labeled p-CMB, it was found that the label was quantitatively released from p-CMB–C′3 following treatment with iodine, as described above for C′2. How oxidation of C′2 can effect not only the active site but also the stability of the C′4,2a complex is presently not understood.

F.   ATTACHMENT OF C′3

Following the assembly of the C′4,2a complex on the cell surface, the next step in immune hemolysis is the binding of the third component. The accumulated evidence indicates that the mechanism of binding of C′3 follows the principle of C′2 and C′4 fixation. The C′3 molecule is activated by an activating enzyme, which in this case is the cell-bound C′4,2a complex, and is thereby enabled to react with suitable receptor groups on the cell membrane (Müller-Eberhard et al., 1966a). The state of activation is apparently of very short duration and with its termination a C′3 molecule that has not become bound loses the capacity of fixation and remains as hemolytically inactive C'3 (C'3i) in the fluid phase. The state of activation is inferred from the fact that neither native C'3 nor C'3i are capable of reacting with C'3 receptor groups. Since this ability is acquired by the native molecule through the action of the C'4,2a complex, which also renders it inactive, conversion of C'3 to C'3i is assumed to proceed via the activated form, C'3a. The chemical equivalent of the state of activation is still unknown.

The kinetics of the C'3-binding reaction as well as its temperature and pH dependence are in accordance with its enzymatic nature. Specific uptake has a temperature optimum of 32°C. and a pH optimum of 6.6. Binding requires that the responsible enzyme, C'4,2a, is itself cell-bound; the fluid phase version, (C'4i,2)a, although effecting conversion is unable to catalyze C'3 binding to cells. The efficiency of C'3 binding to EAC'1a,4,2a varies between 5 and 15% in terms of C'3 input as was determined with C'3-^{125}I; the remaining 85–95% appear as C'3i in the fluid phase. By introducing differential radioactive labels into C'3 and C'4 it was possible to determine that a single C'4,2a site can catalyze binding of several hundred C'3 molecules. It was, therefore, postulated that the site of binding was distinct and spacially separated from the site of activation. The different distribution of C'3 and C'4 on the cell surface could be verified by electron microscopy using ferritin conjugates of anti-C'3 and anti-C'4. Sections of EAC'1a,4,2a,3 cells revealed only a few C'4 sites along the circumference of a cell, but many C'3 molecules which in some instances appeared to be evenly distributed. Limiting the number of C'3 molecules and examining the surface of the cells by means of the grid technique, C'3 appeared to have a patchy distribution. The possibility was considered that each patch of C'3 molecules was oriented around a C'4,2a site (Mardiney et al., 1968; Müller-Eberhard et al., 1966c). Little is known regarding the chemical nature of the binding site of C'3 on the cell surface. Antibody molecules have been considered as possible receptors. Leddy et al. (1965), working with the human cold agglutinin system, found that immunochemically detectable C'3 (and C'4) was carried along with the antibody when the latter was transferred from one cell batch to another. On the other hand, it has been shown that C'3a can attach directly to membrane receptor sites, as EC'4,2a, i.e., cells that are entirely devoid of antibody were just as efficient in taking up C'3 as EAC'4,2a which contained eight units of rabbit hemolysin (Müller-Eberhard et al., 1966a). That C'3 is physically attached to the cell surface during immune hemolytic reaction was evident even before radioactively labeled components were available. EAC'1a,4,2a,3 is agglutinated by anti-C'3 (Harboe et al., 1963;

Wellensiek *et al.*, 1963), and anti-C'3 can block the hemolytic reactivity of cell-bound C'3 (Wellensiek *et al.*, 1963).

Dissociation of cell-bound C'3 was effected by 0.0017 M cholate–deoxycholate at pH 7.4, or by exposure of the cells to 37°C. for 3 hours (Dalmasso and Müller-Eberhard, 1967). Both methods yielded material which was comparable to C'3i in ultracentrifugal and electrophoretic properties. Thus, it is evident that C'3a is bound to cell membranes in macromolecular form, although it is not known at present whether the low molecular weight C'3 fragment which is produced by the activating enzyme remains loosely associated with the bound C'3 molecule or whether it is released in the process of fixation.

Although a C'4,2a site has the capacity of catalyzing fixation of large numbers of C'3 molecules, the requirements in immune hemolysis are much more modest. At 63% lysis only a few C'3 molecules per site become fixed; the exact number varies inversely with the number of C'4,2a sites per cell. If this number is very large, i.e., greater than one thousand, an average of one bound C'3 molecule per site may be sufficient to commit the cell to the lytic action of the subsequent complement components (Müller-Eberhard *et al.*, 1966a). It appears probable that for the formation of a potentially lytic C'3 site a C'3 molecule has to be placed in a critical position on the cell surface.

G. FORMATION OF A STABLE INTERMEDIATE COMPLEX BY THE ACTION OF C'5, C'6, AND C'7

1. Formation of Intermediate Complexes at Low Ionic Strength

This phase of the immune hemolytic reaction is less well understood than the preceding steps. This is in part due to the difficulty encountered in preparing intermediate products that contain active C'5 sites or active C'6 sites. By contrast, the complex EAC'1a,4,2a,3,5,6,7 is readily produced when EAC'1a,4,2a,3 is reacted with a mixture of C'5, C'6, and C'7. These difficulties of preparing two of the three theoretically possible intermediate products of this reaction phase may not entirely be of a technical nature. The possibility exists that they reflect an essential feature of the mode of action of C'5, C'6, and C'7. Information presently available has been accumulated essentially in three laboratories and the results obtained by the three teams are roughly in agreement. A critical review of the data permits the formulation of two distinct hypotheses regarding the mechanism of action of C'5, C'6, and C'7 which may be referred to as the "sequential action" and "functional unit" hypotheses.

The work of Inoue and Nelson (1965, 1966), who employed functionally pure components from guinea pig serum, may be summarized as follows. Intermediate complexes can actually be produced which contain active C'3b (C'5) sites or active C'3e (C'6) sites; the trick is to use buffers of low ionic strength ($\mu = 0.074$). They found that, although the reactivity of C'3b and C'3e is virtually nil at physiological ionic strength, it increases sharply as the ionic strength is lowered and is still rising at $\mu = 0.03$. The reactivity of the C'3b and C'3e sites is lost during subsequent incubation at 30°C. with half-lives of 16 and 75 minutes, respectively. If incubation at 30°C. is carried out at physiological ionic strength, the half-lives of the two intermediate complexes is sharply decreased, suggesting dissocitaion of the components from the cell surface. Accordingly, the efficiency of C'3b, C'3e, or C'3f in forming the respective intermediate complexes was exceedingly low at physiological ionic strength. In contrast, C'3f sites once formed were stable at 30°C. regardless of ionic strength; their half-life was 417 minutes (Rommel and Stolfi, 1966). It is not clear from these studies how immune hemolysis can proceed with high efficiency even at physiological ionic strength, that is, under conditions that are highly unfavorable for the formation of presumably essential intermediate products.

Wellensiek and Klein (1965), also working with guinea pig components, reported that they could prepare an intermediate complex containing C'5 sites and a complex containing C'7 sites. Formation of C'5 sites took place only when a partially purified C'5 preparation was used; highly purified C'5 was not effective. They considered the possibility that their less pure C'5 also contained C'6 and that the "C'5 cells" represented actually one of the later products of the complement reaction. The cells that had reacted with C'7 were found to be considerably more stable than the preceding three intermediate complexes.

2.   Mode of Action and Fate of C'5, C'6, and C'7 at Physiological Ionic Strength

Studies in the author's laboratory (Nilsson and Müller-Eberhard, 1966, 1967b) using human complement components have encountered similar difficulties in preparing intermediate products characterized by the presence of C'5 or C'6 sites. To make these intermediate complexes was virtually impossible at physiological ionic strength and of doubtful success even at $\mu = 0.06$. By contrast, a stable intermediate complex could readily and reproducibly be made when EAC'1a,4,2a,3 was reacted with a mixture of C'5, C'6, and C'7. The question arose, therefore, whether C'5, C'6, and C'7 were interdependent in their action. Kinetic experiments indicated a definite sequence of action, the order of which is expressed in the numerical designation of the three components. In addition, dose–response measurements yielded a sigmoidal curve, which is usually indicative of a multiple-step reaction. In exceptional cases, however, a sigmoidal dose–response curve may arise from dissociation of the active principle into inactive subunits upon dilution. Dissociation of the active principle at low concentration has been cited as explanation for sigmoidal dose–response curves encountered with tryptophan synthetase (Creighton and Yanofsky, 1966), β-glucuronidase (Bernfeld et al., 1954), and hyaluronidase (Bernfeld et al., 1961). Increasing the relative C′5 concentration in dose–response measurements with C′5, C′6, and C′7 resulted in a marked change of the shape of the dose–response curve. It now appeared concave toward the dose scale, and upon transformation to a plot of average number of sites per cell versus relative C′5, C′6, and C′7 concentration, a linear relation was obtained. This finding was consistent with the hypothesis that the three components act as one functional unit.

Formation of the stable intermediate complex (EAC′1a,4,2a,3,5,6,7) from EAC′1a,4,2a,3 by C′5,6,7 was found to be markedly temperature dependent, the optimum being between 20° and 30°C. The reaction was also markedly dependent upon pH with an optimum at pH 6.4. Depletion experiments showed that much of the activity of the three components disappeared from the fluid phase during the reaction—a finding which was also made by Inoue and Nelson (1965, 1966) for guinea pig C′3b, C′3e, and C′3f. The question of whether depletion of these activities is due to destruction in the fluid phase or to uptake of the active principle by the cells was tackled with respect to C′5. As this component is available in highly purified form, it can be labeled with radioactive iodine and its fate in immune hemolysis may thus be traced independent of its hemolytic activity. Initial experiments indicated that C′5 acted without becoming bound to the cell (Nilsson and Müller-Eberhard, 1965). However, when the specific radioactivity was substantially increased, it was found that a very small amount, less than 0.5% of C′5 input, became specifically bound to the red cell membrane. The remaining 99.5% accumulated during the reaction as inactive C′5 (C′5i) in the fluid phase (Nilsson and Müller-Eberhard, 1967b).

3. Physicochemical Interaction of C′5, C′6, and C′7

Evidence for physicochemical interaction between the three components was obtained by zone ultracentrifugation (Nilsson and Müller-Eberhard, 1967b). Although C′6 and C′7 have an s rate of 5 to 6 S when examined in purified form, both activities sedimented at a faster rate (8–10 S) if they were analyzed in mixtures with C′5 in which the concentration of C'5 was relatively large. At low concentrations of C'5, both C'6 and C'7 were found to sediment at intermediate rates. The phenomenon was interpreted to indicate that C'6 and C'7 possess an affinity for C'5 which enables them to enter into reversible complexes with this component. These experiments also showed that association of the three components to a complex depended on the concentration of C'5 and on environmental conditions. Interaction of the three components was also observed upon ultracentrifugation of whole human serum.

4. Two Alternative Hypotheses

Studies so far performed on the reaction mechanism of C'5, C'6, and C'7 may be interpreted in the form of two distinct hypotheses. Both hypotheses recognize that the action of C'5, C'6, and C'7 depends on the presence of two functional sites on the cell surface, the C'4,2a and the C'3 site. Presumably these two sites are required for the activation of the three components, a process which most probably involves also the C'3-dependent peptidase activity (Cooper and Becker, 1967). The alternative interpretation of the data relates to the question of whether C'5, C'6, and C'7 act sequentially or as one functional unit. The "sequential action" hypothesis is primarily substantiated by the results of kinetic experiments; it is compromised by the difficulty to demonstrate intermediate products, particularly at physiological conditions. The "functional unit" hypothesis is chiefly supported by the finding that C'5, C'6, and C'7 are capable of interacting physicochemically. The kinetic data and the sigmoidal shape of the dose–response curve tend to refute the latter hypothesis. The possibility exists, however, that the kinetic data reveal sequence of activation of the three components rather than sequence of action. Furthermore, the sigmoidal dose–response curve need not necessarily indicate a multiple-step reaction but rather progressive dissociation of the C'5,6,7 complex with dilution. If the C'5,6,7 complex would prove to be one functional unit, its mode of action would most likely become explicable in terms of the function of a structurally complex enzyme. Further studies are needed to delineate the precise mechanism of this reaction phase.

H. The Terminal Steps of Immune Cytolysis

1. The C'8 Reaction Step

The terminal events of cell damage by complement involve two components, C'8 and C'9. The reaction of these two components with the thermostable intermediate complex (EAC'1a,4,2a,3,5,6,7) is readily separable into two parts, one involving C′8 and leading to the formation of active C′8 sites on the membrane, and the other involving C′9, which results in the activation of a membrane effector system. From the work of Rommel and Stolfi (1966) it is known that C′8 is depleted in the fluid phase as active C′8 sites are formed on the cell surface. The reaction is temperature dependent. However, it will proceed even at 0°C. if sufficient time is allowed. The C′8 site has a half-life of 368 minutes at 30°C., and it is stable at 0°C. for at least 24 hours. Linscott and Nishioka (1963) reported evidence that these cells possess increased fragility which was believed to indicate that the cell membrane at this stage has sustained definite damage which, however, is not sufficient to cause breakdown of its function. This subthreshold membrane damage becomes apparent on storage of the cells in the course of which they undergo continuous low-grade lysis (Stolfi, 1967). Rommel and Stolfi (1966) made the pertinent observation that C′8 after inactivation of alkaline pH and low ionic strength retains its ability to react with C′7 sites, as it can block conversion of these sites by hemolytically active C′8. The blocked sites are able to deplete C′9 activity from the fluid phase to the same extent as sites prepared with active C′8. The observations suggest that C′8 is combining physically with membrane sites and that C′9 reacts with cell-bound C′8.

Further evidence in support of this view came from studies with anti-human C′8 (Manni and Müller-Eberhard, 1968). The antibody, which was raised in rabbits by immunization with functionally pure C′8, inhibits the hemolytic activity of C′8 and has no effect on C′9. In addition, it reduces the susceptibility of C′8 cells to lysis by C′9 and this fact strongly indicates that C′8 action involves firm binding of this component to the cell surface.

2. The C′9 Reaction Step

Experiments with highly purified human C′9 which was labeled with radioactive iodine have revealed the following mode of action of the component (Hadding and Müller-Eberhard, 1968). During incubation of C′9 at 32°C. with cells containing active C′8 sites, C′9 activity rapidly disappears from the fluid phase. Concomitant with disappearance of C′9 activity, C′9 is specifically taken up by the cells and remains firmly attached to them. The time course of hemolysis parallels that of C′9 binding; however, there is a definite lag phase between onset of binding and onset of lysis. The C′9 reaction step is markedly temperature dependent. Electron microscopy revealed the appearance of the characteristic ultrastructural lesions following C′9 action. No such lesions could be detected with certainty in membranes of cells containing C'8 sites (Hadding *et al.*, 1966).

3. The Effect of EDTA and Phenanthroline

Frank *et al.* (1964) reported the ability of 0.09 M EDTA to inhibit immune hemolysis subsequent to the C'2 step. They also provided evidence that 0.09 M EDTA interferes with E* transformation, i.e., conversion of a cell that has reacted with complement to a damaged cell. Following up their observations it was found that C'9 can exert its effect in the presence of a high concentration of EDTA (Hadding and Müller-Eberhard, 1968). When C'8 cells are treated with C'9 in the presence of 0.09 M EDTA, no lysis occurs; however, hemolysis ensues without lag following removal of the fluid phase and resuspension of the cells in buffer devoid of EDTA and C'9. The low-level lysis of C'8 cells has also been found to be inhibited by 0.09 M EDTA. The data are consistent with the view that 0.09 M EDTA interferes with a reaction occurring within the membrane which is triggered by C'9 and which, in turn, causes the critical modification of the membrane that becomes manifest in cell lysis.

In exploring whether the chelating function of EDTA was responsible for its inhibitory effect, phenanthroline, another chelating agent, was tested (Hadding and Müller-Eberhard, 1968). This compound not only failed to inhibit E* transformation, but apparently enhanced C'9 action. In addition, phenanthroline at a concentration of 0.01 M was found to be able to substitute C'9 in that it could initiate lysis of C'8 cells but not of C'7 cells. The rate and extent of lysis of C'8 cells increased with the number of C'8 sites per cell. The mode of action of phenanthroline is unknown, and yet it appears likely that it affects cell-bound C'8.

Although the data accumulated to date are insufficient to give a clear picture of the terminal events in immune hemolysis, they permit formulation of the following working hypothesis. It proposes that C'8 attaches itself to a site on the cell surface which was prepared by the action of C'5, C'6, and C'7. Then C'9 is bound to C'8 which becomes fully activated in the process. Then C'8,9 acts on a membrane constituent tentatively designated the precursor of a membrane effector. The active site is assumed to reside in C'8, since C'8 cells exhibit a certain degree of leakiness and since C'9 action can be mimicked by phenanthroline. Action of C'8,9 on the membrane effector leads to its activation. The actual damage is subsequently caused by this membrane constituent, which can be inhibited by 0.09 M EDTA.

The chemical nature of the process that causes the complement defect is still unknown. Fischer and Haupt (1961, 1965) several years ago invoked lysolecithin as the actual agent responsible for membrane damage by complement. They speculated that one of the terminal complement components upon activation assumes the activity of lecithinase A and liberates lysolecithin by acting on a lecithin-containing complement component. This concept presupposed a phospholipid-containing serum constituent as a necessary member of the complement system. It could be shown, however, with considerable certainty that lipids or lipoproteins were not essential for a hemolytically active complement system (Dalmasso and Müller-Eberhard, 1966). Lately, Fischer (1965) considers the possibility that endogenous lysolecithin derived from membrane phospholipids accumulates in the membrane as a result of inhibition of "safety enzymes" by complement. Although there is no evidence at present for lysolecithin playing any role in complement-dependent cytolysis, the participation of membrane-associated enzymes seems an intriguing possibility.

VI.   Enzymatic Activities Associated with the Complement System

A. C'1 ESTERASE

In 1955 Levine (1955) showed that EAC'1a,4 can be inactivated by treatment with DFP. Independently, Becker (1955) found the DFP effect on complement and reported a year later (Becker, 1956a,b) that it is the C'1a activity of EAC'1a,4 that is affected by DFP. C'1 in serum was completely resistant to inhibition; however, it became susceptible in the presence of sensitized sheep erythrocytes. Becker postulated that C'1 exists in serum as a proenzyme which is resistant to DFP. Upon interaction with sensitized cells it becomes activated and thus susceptible to inhibition by DFP. The esterase activity of C'1 could be demonstrated directly by its ability to hydrolyze TAMe and N-acetyl-L-tyrosine ethyl ester (ATE). The substrate TAMe could protect C'1a activity on sensitized cells against inhibition by DFP. At the same time, Lepow and his associates (Lepow et al., 1956a,b; Ratnoff and Lepow, 1957; Lepow et al., 1958) discovered esterase activity to be associated with complement-treated immune precipitates and they attributed the enzyme activity to the active site of the first component. As delineated above, it has been demonstrated by Lepow et al. (1963; Haines and Lepow, 1964b) that the estrolytic site of the first component is resident in the C'1s subunit. The natural substrates of the enzyme are C'2 and C'4, both of which are proteins, and, although the nature of the bonds attacked in these two components is still unknown, the possibility has to be considered that C'1 esterase possesses peptidase activity.

B. C'4,2a

The enzyme nature of this complex was originally inferred from the quantitation of specific uptake of radioactively labeled C'3 by cells containing C'2a on their surface (Müller-Eberhard *et al.*, 1966a). One C'4,2a site was found to be instrumental in the fixation of several hundred C'3 molecules within a few minutes at 37°C. provided C'3 was present in excess. The reaction was found to be pH and temperature dependent and to follow enzyme kinetics. When the C'4,2a complex could be prepared in cellfree solution (Müller-Eberhard *et al.*, 1967a), its effect on C'3 could be studied directly. A given amount of the complex was found to convert several thousand times this amount of C'3 to C'3i within 20 minutes at 37°C. C'3 undergoes cleavage in the course of this reaction and a low molecular weight fragment becomes dissociable (Dias Da Silva and Lepow, 1967; Müller-Eberhard *et al.*, 1967a). The nature of the bond in the C'3 molecule that is cleaved by C'4,2a has not yet been determined. Similarly, studies are still lacking in which the effect of the enzyme on synthetic substrates is examined.

C. C'3-DEPENDENT PEPTIDASE ACTIVITY

It has been known for some time that peptides containing an aromatic amino acid inhibit the action of guinea pig complement in immune hemolysis (Cushman *et al.*, 1957; Plescia *et al.*, 1957). This inhibitory effect was considered to be indicative of the presence of a peptidase in complement which is essential for the system's hemolytic activity. Basch (1965) showed that inhibition by aromatic amino acids took place between the EAC'1a,4,2a step and the intermediate that has reacted with C'5, C'6, and C'7. He also showed that the immune adherence phenomenon was inhibited by these compounds, thereby supplying suggestive evidence that cell-bound C'3 was the affected component.

Lately, Cooper and Becker (1967) demonstrated peptidase activity to be associated with EAC'1a,4,2a,3 which appeared to be a function of cell-bound C'3 since EAC'1a,4,2a lacked this activity. They found that a number of aromatic peptides, including glycyl-L-tyrosine, were hydrolyzed by the enzyme and that hydrolysis showed zero-order kinetics and a bell-shaped pH activity curve with an optimum at pH 7.6 to 7.8. The Michaelis constant, K_m, was $1.13 \times 10^{-3} M$. They encountered some circumstances where C'3 activity on the cells was high and peptidase activity low or absent. When the intermediate complex was prepared with partially purified C'3 the discrepancy between C'3 activity and enzyme activity became more marked. Participation of later acting components in the generation of peptidase activity was excluded by showing that C'5 through C'9 had no effect on the enzyme activity. It was possible, however, to detect in' serum a low molecular weight, dialyzable factor which was able to enhance the hemolytic, immune adherence and peptidase activities of C'3 on EAC'1a,4,2a,3. The nature of this factor has not been determined (Cooper, 1967). Further analysis with purified human complement components showed that measurable hydrolysis required five effective molecules of oxidized C'2, less than 100 bound molecules of C'3-^{125}I, and 300–400 bound molecules of C'4-^{125}I per cell. Hydrolysis was dependent on C'3 up to approximately 100 bound molecules per cell. The ratio of bound C'3 to bound C'4, necessary to achieve hydrolysis, was, therefore, less than 1. It was concluded that generation of the enzymatic activity may depend on a critical spacial relation of C'3 to a C'4,2a site (Cooper and Müller-Eberhard, 1967a). The natural substrate of the enzyme has not been defined; however, it is likely to be at least one of the three components, C'5,6,7.

VII. The One-Hit Theory of Immune Hemolysis and the Problem of Quantitation of Complement Components

A. DEVELOPMENT AND FORMULATION OF THE ONE-HIT THEORY

Whereas the biologist may attribute little significance to the question of whether immune hemolysis is a "single-hit" or a "multiple-hit" process, this question becomes critical when the components of complement are to be quantitated in absolute terms on the basis of their hemolytic activity. If one hit, i.e., one complete complement reaction occurring at a single site, should suffice to lyse a cell, far less complement would be required for lysis than if lysis depended on an accumulation of hits. In other words, an absolute quantitation of complement by activity measurements becomes feasible and meaningful only if the fundamental mechanism of immune hemolysis is known. This must include detailed information of the reaction mechanisms of the different components as their efficiency may differ widely. Certainly, the ideal in biochemistry remains quantitation of an active principle on an absolute weight basis, and this should be the ultimate aim in designing methods for the quantitation of complement components. As this cannot be accomplished conveniently in a direct fashion for all nine complement components, an indirect approach is necessary. The prerequisite, however, is that two factors be known—the number of effective hits required for lysis of a cell and the efficiency with which one hit is produced by a given complement component. Thus, the problem of complement quantitation becomes intimately interwoven with considerations of the theory of immune hemolysis.

An earlier theory of immune hemolysis was based upon a mathematical evaluation of the dose–response curve describing the lysis of sensitized sheep erythrocytes by complement. As shown by Leschly (1914), the shape of this curve is sigmoidal; and its mathematical formulation, which was derived empirically by von Krogh (1916), is a probability function. This was interpreted to indicate that immune hemolysis is a multiple-hit process, and Alberty and Baldwin (1951) deduced that approximately ten sites on the surface of a cell must react with antibody and complement for lysis to occur.

The multiple-hit theory was criticized by Mayer (1961a), who several years ago introduced the one-hit or noncumulative theory. It is based on the observation that, under certain experimental conditions, the absolute velocity and extent of the lytic reaction are independent of the total number of cells in the reaction mixture. These findings were, according to Mayer, "entirely in accord with a one-hit reaction mechanism in the sense that a single lesion, S^*, suffices for lysis and that a single molecule of at least one of the complement components at some stage of the reaction sequence suffices for production of this single lesion." If it was a one-hit process, the dose–response curve should be concave toward the dose scale instead of being sigmoidal. The deviation of the experimental from the theoretical curve was explained on the basis of the observed exponential rate dependence of E^* formation (Rapp, 1958) and the decay of EAC'1a,4,2a (Mayer et al., 1954). According to Mayer (1965) the validity of the one-hit concept was demonstrated by showing that the experimental titration data for C'1 (Hoffman, 1960; Borsos and Rapp, 1965c), C'2 (Borsos et al., 1961b) and C'4 (Hoffman, 1960) fit the theoretical response curves calculated from the binomial probability distribution for the threshold value $r = 1$ (one-hit curve).

B. Effective and Potentially Effective Molecules

For the quantitative analysis of dose–response curves, Mayer (1961a) applied the Poisson distribution function which yields the relationship $z = -\ln (1 - y)$, provided one lesion, S^*, is sufficient for cell lysis, i.e., the average number of S^* per cell equals the negative, natural logarithm of the surviving cells. At 63% lysis ($y = 0.63$) the z value is unity ($z = 1$). Thus, 63% conversion of a suspension of EAC'1a to EAC'1a,4 implies that a successful reaction of an average of one C'4 molecule per cell has occurred. The total number of molecules present in a given sample is obtained according to this approach by multiplying the number of cells by the reciprocal of the relative dilution of the sample at which $-\ln (1-y) = 1$. According to Mayer (1961a), such absolute molecular titrations represent minimal estimates as they do not include determination of that portion of complement that was expended in unfruitful reactions. Recognizing this limitation the results are expressed in terms of "effective molecules."

Application of this method for quantitation of C'4 in guinea pig serum yielded a value of 3–4 × 10^{12} effective molecules per milliliter serum (Mayer, 1961b). For human serum, a value of 4 × 10^{12} effective molecules per milliliter was obtained (Cooper and Müller-Eberhard, 1967b). In terms of weight, this value corresponds approximately to 1.5 µg./ml., assuming a molecular weight for C'4 of 230,000. This amount is less than can actually be isolated from human serum in the form of a highly purified, physicochemically and immunologically homogeneous protein. Furthermore, isolation of the protein involves considerable loss, the yield being below 25%. Thus, there appeared to be a marked discrepancy between results obtained by titration and those based on chemical analyses.

The answer to this problem was sought in the characteristics of the C'4 reaction mechanism. As was pointed out above, fixation of C'4 to cell surface acceptors requires activation of the molecule and is accompanied by decay of a large proportion of molecules to unbound, hemolytically inactive C'4i. Although C'4i molecules are, in all probability, derived from potentially effective C'4 molecules, they do not register in the effective molecule titration. It was further considered possible that not all of the specifically bound C'4 molecules are placed in positions where they are hemolytically effective. Some bound C'4 molecules, though capable of reacting with C'2, may be rendered ineffective due to resistance of their immediate environment on the cell surface to damage by complement. Also these molecules would escape detection in the titration procedure.

Effective molecule titrations were performed on highly purified human C'4 which was labeled with radioactive iodine to permit an independent and direct tracing of its fate. Under the experimental conditions employed, the efficiency of uptake was approximately 6%, i.e., 6% of the molecules were specifically bound, whereas 94% was converted to C'4i. The efficiency of the specifically bound C'4 was approximately 5%, i.e., an average of 20 C'4 molecules per cell had become specifically bound at 63% conversion of EAC′1a to EAC′1a,4, corresponding to $-\ln(1-y)$ $= 1$. It, therefore, follows that in this particular experiment a correction factor of 330 has to be applied to convert the number of effective molecules to the absolute number of C′4 molecules in the reaction mixture. Thus, the true amount of C′4 in human serum according to these measurements would be $4 \times 10^{12} \times 330$ molecules per milliliter or 495 μg./ml. (Cooper and Müller-Eberhard, 1967b).

With the availability of monospecific antisera to several human complement components it became possible to determine the serum concentration of the respective components immunochemically. Using the single radial diffusion technique and known amounts of isolated complement components as standards, the following mean values were obtained by analysis of ten different sera: C′3, 1200 ± 300 μg./ml.; C′4, 430 ± 120 μg./ml.; and C′5, 75 ± 24 μg./ml. (Kohler and Müller-Eberhard, 1967). It is apparent that the immunochemical determination of C′4 is in excellent agreement with the value derived from effective molecule titrations after it was corrected for the efficiency of binding of C′4 and the hemolytic efficiency of bound C′4. Similar correction factors are presently being determined for some other components. Preliminary results indicate that the hemolytic efficiency of C′1 is greater than that of C′4 and C′2 and that the efficiency of C′3 is relatively low.

With the development of the one-hit theory, on the one hand, and of the chemical approach to an understanding of complement, on the other, observers of the complement field have become increasingly uncertain of validity and meaning of quantitative estimates of complement components. At times the disparity of results obtained by different schools of thought has been marked. Those interested in quantification by use of the one-hit theory arrived at values which, by virtue of the theoretical concept employed, tended to be relatively low. Those pursuing quantitation according to the principles of protein chemistry encountered values that in some instances were comparatively high. At least for one complement component, namely C′4, the discrepancy has now been explained through an analysis of the mode of action of this component and the determination of its hemolytic efficiency. By tracing the fate of C′4 molecules in the immune hemolytic reaction using a radioactive label, it has come to light that only 1 out of 15 to 20 molecules succeeds in establishing a cell-bound C′4 site; the other molecules undergo inactivation in the process. In addition, only 1 out of 20 to 30 cell-bound C′4 molecules registers as an "effective molecule" in the "effective molecule titration." The reason for this is unclear; it is conceivable, however, that a C′4 molecule has to be placed in a critical position on the cell surface to become hemolytically effective. Thus, 300–400 potentially effective C′4 molecules are required to produce one effective molecule in the sense of the one-hit theory. This should not come as a surprise in view of Mayer's cautious advice that molecular titrations based on the one-hit theory must be regarded as minimal estimates because of the likelihood of unfruitful reactions. He also pointed out that proper correction factors will have to be determined. For human C′4 the correction factor has now been obtained. Neglecting it would be just as detrimental to efforts of quantitation as being uncertain about whether immune hemolysis is a one-hit or a multiple-hit process. It should be emphasized at this point that it is not possible to extrapolate from the correction factor obtained for C′4 to similar factors of other components. These will have to be worked out for each component separately, according to the procedure that has been established for C′4.

VIII.   Mechanism of Lysis of Paroxysmal Nocturnal Hemoglobinuria Erythrocytes

The hemolytic reaction in paroxysmal nocturnal hemoglobinuria (PNH) has been discussed recently by Hinz (1966), and the interested reader is referred to his comprehensive review for detailed information. The problem will be treated here only inasmuch as it relates to the important concept of induction of hemolysis from the fluid phase. As pointed out above, complement can react with erythrocytes without the participation of antibody. Such reactions may result in frank hemolysis, although the efficiency of complement in antibody-free systems is usually low. It was shown that C′4 may attach to nonsensitized sheep erythrocytes through the action of C′1 esterase in the fluid phase giving rise to the formation of EC′4 (Müller-Eberhard and Lepow, 1965). Similarly, C′2 may become attached and activated by fluid phase C′1 esterase to form EC′4,2a (Müller-Eberhard et al., 1967a). Both C′3 and C′4 became cell-bound, and lysis ensued when nonsensitized cells were incubated with autologous complement in the presence of polyethylene glycol which acts as an activator of C′1 (Dalmasso and Müller-Eberhard, 1964). Finally, direct action of C′5, C′6, and C′7 on E could be demonstrated by treating E with a preincubated mixture of C′1 esterase, C′2, C′3, C′4, C′5, C′6, and C′7. Preincubation undoubtedly led to the formation of (C′4i,2)a and to conversion of C′3, thereby rendering these three components unable to combine with the cells when they were later added to the reaction mixture. Following a period of incubation and subsequent washing, a certain proportion of the cells behaved with respect to EDTA–serum or C′8 and C′9 as EC′5,6,7 (Müller-Eberhard et al., 1966d).

Paroxysmal nocturnal hemoglobinuria erythrocytes may be lysed *in vitro* without participation of antibody (Hinz, 1966). The reaction requires a high serum concentration, preferably undiluted serum, slightly acid pH (pH optimum 6.5–6.8) (Hinz, 1966; Yachnin and Ruthenberg, 1965a), and magnesium but not calcium ions (Hinz *et al.*, 1956; Yachnin and Ruthenberg, 1964). It is enhanced by C′1 esterase and by activators of C′1 such as polyinosinic acid and dextran sulfate (Yachnin and Ruthenberg, 1965b). There is overwhelming evidence indicating that lysis of PNH erythrocytes *in vitro* is caused by complement. It does not seem to differ from classic immune hemolysis except, as was pointed out by Yachnin (1965), in the localization of the early steps of the complement reaction. In antibody-mediated immune hemolysis, C′1 is activated upon combination with antibody and it is primarily on the cell surface that the complement reaction is triggered. In PNH hemolysis the early steps of the reaction appear to take place in the fluid phase. Apparently a small amount of activated C′1 esterase in the serum is sufficient to set in motion the chain of events that culminates in lysis of these abnormal cells. Perhaps fluid-phase C′1 esterase effects the assembly of the C′4,2a complex on the surface of the PNH erythrocytes which then causes binding of C′3 and so forth. Jenkins *et al.* (1966) could demonstrate immunochemically detectable C′3 and C′4 on unlysed cells after the acid hemolysis reaction in serum. On the other hand, it is conceivable that the PNH cells are attacked by complement at a later stage of the complement reaction. The C′4,2a complex may be formed in the fluid phase and may act here on subsequent components activating them for their action on the PNH cells. That this is a feasible mechanism was demonstrated in two different ways. First, Yachnin (1965, 1966) rendered PNH cells susceptible to lysis in EDTA–serum by treating them with a partially purified preparation of later acting components, probably activated C′5, C′6, and C′7. Second, PNH cells could be lysed in EDTA–serum by addition of preformed (C′4i,2)a (Müller-Eberhard *et al.*, 1967a), which acts on C′3 and later components but is unable to attach itself or C′3 to the cell surface. Thus, these experiments indicate that the attack of complement on PNH cells may start at a stage as late as the C′5,6,7 step. Again, the susceptibility of PNH cells to lysis by complement from the fluid phase is not peculiar to these cells; the phenomenon can be observed also with normal human (Yachnin, 1965) or sheep erythrocytes. But PNH cells are approximately 30 times as sensitive to complement action as normal cells (Rosse and Dacie, 1966) and for this reason lysis by fluid-phase complement may play a much greater role in PNH than in normal situations.

IX. Effect of Complement on Bacteria

Complement was discovered by its ability to kill bacteria (Buchner, 1889; Bordet and Gengou, 1901); although the initial observations were made some 70 years ago, the exact mechanism of the bactericidal reaction has not been worked out thus far. The problem is complicated by the fact that bacteria possess not only a cell membrane like mammalian cells but also a cell wall and, in certain cases, even an additional capsule. It was postulated by Muschel (1965) that the bactericidal reaction of complement results from damage to the bacterial cell membrane. As the membrane is covered by the cell wall it is believed that the complement-activating, antigen–antibody reaction occurs on the wall of the organism. This concept implies that the activated complement components must transfer from the site of activation through the wall to the site of action on the cell membrane. Direct evidence for such a mechanism is lacking; however, characteristic ultrastructural lesions have been found in the membrane of *Escherichia coli* after treatment of this organism with antibody and complement (Bladen *et al.*, 1966). The lesions measure 80–100 Å. in diameter, are surrounded by a less dense ring, and resemble in appearance and size those observed in complement-damaged erythrocyte membranes (Humphrey and Dourmashkin, 1965).

Whether a microorganism is susceptible or resistant to the bactericidal effect of complement appears to be determined by the properties of its cell wall or capsule. In general, gram-negative organisms are susceptible and gram-positive species are resistant. The differential susceptibility has been linked to differences in phospholipid content (Wardlaw, 1960) and in thickness (Muschel and Jackson, 1966) of the cell wall. The complement-resistant gram-positive species by comparison possess a wall of greater thickness and lower lipid content. Working with cell-wall-free organisms, i.e., with spheroplasts of the gram-negative species, *E. coli*, and with protoplasts of the gram-positive *Bacillus subtilis*, Muschel and Jackson (1966) demonstrated clearly that the membranes of both types of organisms are affected by antibody and complement, and that the complement resistance of gram-positive bacteria must be due to their particular cell wall. Sensitivity to complement may vary also among the gram-negative species. For *Paracolobactrum ballerup*, a gram-negative organism which was considered to be completely resistant to complement action, Osawa and Muschel (1964) showed that it can be converted to complement sensitivity by cultivation at temperatures above 37°C.

Upon microscopic examination, bacteria killed by antibody and complement do not exhibit any structural distortion (Muschel *et al.*, 1959). Lysis may, however, be produced by subsequent addition of lysozyme which acts on the mucopeptide backbone of the cell wall (Amano *et al.*, 1954; Muschel *et al.*, 1959; Wardlaw, 1962).

X.   Mechanism of Action of Complement in Noncytolytic Reactions

A.   FORMATION OF ANAPHYLATOXIN

1.   Definition of Anaphylatoxin

Anaphylatoxin is a substance of relatively low molecular weight which is able to release histamine from mast cells and which thereby causes smooth muscle contraction and changes in capillary permeability. Systemically, it may cause anaphylactic shock. Anaphylatoxin is quantitated by measuring the contraction of a segment of guinea pig ileum suspended in an organ bath containing Tyrode solution. Its effect on smooth muscles and capillaries is inhibitable by antihistamines. Anaphylatoxin was originally described by Friedberger (1910) at the beginning of this century, who demonstrated that treatment of fresh serum with immune precipitates results in the formation of a substance that elicits anaphylactic shock in animals. He speculated that this substance is derived from the complement system. Osler *et al.* (1959) provided suggestive evidence that anaphylatoxin formation was complement dependent and that its mechanism involved more than the first three components. More recently, Ratnoff and Lepow (1963) found that purified C'1 esterase increased vascular permeability in guinea pig skin and that this effect was inhibited by antihistaminic agents. Furthermore, treatment of guinea pig or rat serum with C'1 esterase produced an agent capable of contracting guinea pig ileum, degranulating mast cells, and releasing histamine from guinea pig tissue (Dias Da Silva and Lepow, 1966b). At this point, there is little doubt about complement being operative in anaphylatoxin generation; the question is how it is produced and from which component.

2.   Formation of Anaphylatoxin from Human C'3

Dias Da Silva and Lepow (1966a) found that interaction of purified human C'1 esterase, C'2, C'3, and C'4 in the presence of magnesium ions resulted in rapid generation of anaphylatoxin activity. Deletion of any one of these factors yielded a biologically inactive system. A possible requirement for C'5 could not be rigorously excluded on the basis of these experiments, since the preparations of C'1 esterase and of C'2 contained small amounts of contaminating C'5. The free solution system was further explored using radioactively labeled, purified C'3 and C'5 (Dias Da Silva and Lepow, 1967; Müller-Eberhard et al., 1967a; Cochrane and Müller-Eberhard, 1967a,b). It was found that in the course of the reaction a fragment is cleaved from the C'3 molecule which could be isolated by passing the reaction mixture through a Biogel 150 column at pH 3 or 5. As indicated by the distribution of the radioactive label, approximately 5% of C'3 was retarded and eluted from the column together with marker substances of molecular weight 8000–12,000. Anaphylatoxin activity corresponded in distribution to the low molecular weight C'3 derived material. A similar fragmentation could not be observed with radioactively labeled C'5. Nor did the addition of increasing amounts of C'5 to the above described reaction mixture lead to an increase in anaphylatoxin production. It was concluded that a histamine-releasing, low molecular weight substance resembling anaphylatoxin is derived from human C'3 by the action of the C'4,2a complex. The active fragment was designated F(a)C'3.

Similar observations were made with radioactively labeled C'3 in a completely different free solution system. Snake venom has been known for a long time to liberate anaphylatoxin from serum (Friedberger et al., 1913; Hahn and Lange, 1956). The responsible factor was isolated from cobra venom (see below) and utilized together with a partially purified β-globulin of human serum to inactivate C'3, which was cleaved in the course of the reaction (Müller-Eberhard et al., 1966c; Müller-Eberhard, 1967b). A small fragment resembling F(a)C'3 was found to exhibit anaphylatoxin-like activity (Cochrane and Müller-Eberhard, 1967a,b). Such fragment or activity could not be derived from human C'5 by subjecting it to an identical treatment. It is not known whether the cobra-factor-produced C'3 fragment is identical with the split off by C'4,2a. Both fragments have in common a similarly low molecular weight as well as the tendency to remain associated with the high molecular weight C'3 fragment, a tendency which can be overcome by an increase in hydrogen ion concentration. After dissociation the low molecular weight material tends to aggregate. These aggregates seem to be biologically inactive.

As mentioned above, brief treatment with trypsin also results in cleavage of F(a) from C'3. Recently, it was possible to show that this trypsin-produced F(a) fragment also possesses anaphylatoxin activity. To demonstrate the activity it is essential to limit the action of trypsin to 1 minute at 20°C., which is sufficient for complete liberation of the fragment. Longer exposure to trypsin causes inactivation of F(a) due to enzymatic degradation (Bokisch *et al.*, 1968).

3. Formation of Anaphylatoxin from Guinea Pig and Human C'5

Working with functionally purified guinea pig components, Jensen (1967) explored the conditions for the formation of anaphylatoxin in a variety of systems. He was unable to demonstrate anaphylatoxin activity in the supernatant of EAC'1a,4,2a treated with C'3. However, upon addition of C'5, biological activity was generated. Destruction of the hemolytic activity of C'5 by heating at 80°C. also destroyed its capacity to generate anaphylatoxin. Although cobra venom factor in conjunction with an unidentified serum cofactor inactivated guinea pig C'3, it did not liberate anaphylatoxin from this component; however, when the two factors were incubated with C'5, activity appeared. Activity could also be produced by treating C'5 with trypsin for 10 minutes at 25°C. Even heated C'5, which was inactive in the immune hemolytic system, could serve as donor of anaphylatoxin when trypsin was used. Jensen concluded that C'5 constitutes the anaphylatoxinogen of guinea pig serum.

Using highly purified human C'5, anaphylatoxin activity could be generated by treatment with trypsin for 5 minutes at 37°C. (Cochrane and Müller-Eberhard, 1967a). The activity was found to reside in a fragment of the C'5 molecule [F(a)C'5] which has a molecular weight of approximately 10,000. Smooth muscle contraction brought about by F(a)C'5 was inhibited with antihistamine. Anaphylatoxin activity could also be derived from human C'5 by the action of EAC'4,2a,3, particularly if oxidized C'2 was used in preparing the complex.

These studies indicate that both C'3 and C'5 may serve as parent molecules of anaphylatoxin. The two active principles are quite distinct, however, since it was found that human F(a)C'3 failed to cross-desensitize the guinea pig ileum to the contracting capacity of human F(a)C'5 and vice versa (Cochrane and Müller-Eberhard, 1967a).

4. Chemical Properties of Anaphylatoxin

Anaphylatoxin has been isolated from rat and pig serum after generation of the activity by treatment of these sera with cobra factor or dextran (Stegemann *et al.*, 1964a). Activity was detected with amounts as small as 1 μg.; but usually 10–15 μg. were required to produce an effect on guinea pig ileum. The substance was inactivated by subtilopeptidase A, chymotrypsin A, trypsin, and carboxypeptidase A and B. It was most stable in acid milieu, it was not dialyzable, and yet on Sephadex filtration it behaved like a substance of molecular weight of 2000 to 4000. The molecule was of basic character, being strongly adsorbed to carboxymethylcellulose and not at all adsorbed to DEAE-cellulose. Molecular weight determinations in the analytical ultracentrifuge yielded a value of 29,000 which is considerably higher than the estimate based on Sephadex filtration. The possibility was considered that the latter value might be too low due to interaction of the basic material with the gel. Or the ultracentrifugal value might be too high due to aggregation of the material. By Edman degradation and following dinitrophenylation, the amino terminal residue of rat and pig anaphylatoxin was found to be arginine. Although both carboxypeptidases inactivated anaphylatoxin, they did not liberate a carboxy terminal residue. Amide nitrogen determination revealed that almost all carboxyl groups were substituted (Stegemann *et al.*, 1964b). Fingerprinting after tryptic digestion disclosed thirteen distinct spots. The molecule contained mannose and another unidentified carbohydrate (Stegemann *et al.*, 1965).

B. FORMATION OF LEUKOCYTE CHEMOTACTIC FACTORS

1. The Macromolecular Factor

Studying the chemotactic attraction of polymorphonuclear leukocytes by antigen–antibody complexes *in vitro*, Boyden (1962) found that the reaction was mediated by a serum factor. He implicated the complement system since heating of serum and EDTA prevented the generation of the activity. These initial studies were extended by Ward *et al.* (1965, 1966) who utilized defined red cell–antibody–complement complexes. They found that cells in the state EAC'1a,4,2a,3 did not give rise to chemotactic migration of leukocytes; however, addition of C'5, C'6 and C'7 resulted in generation of activity. C'5, C'6, and C'7 are known to interact in cellfree solution to form reversible complexes (Nilsson and Müller-Eberhard, 1967b). Similarly, the chemotactic activity was found to sediment on zone ultracentrifugation together with the C'5,6,7 complex when conditions were employed favoring association of these components. At environmental conditions that led to dissociation of the complex and separation of the components during ultracentrifugation the chemotactic activity disappeared. Since the process was reversible, it was concluded that the chemotactic activity resided in the C'5,6,7 complex. Both C'5 and a mixture of C'6 and C'7 had to undergo activation by treatment with EAC'1a,4,2a,3 before they became active in the chemotactic system. The native components lacked this activity. Activation could be inhibited by N-CBZ-glycyl-L-phenylalanine and by N-CBZα-glutamyl-L-tyrosine in amounts that are known to inhibit immune hemolysis and immune adherence (Basch, 1965). The activated chemotactic factor could be inhibited by glutamyl-L-tyrosine, but not by glycyl-L-phenylalanine. The inhibition was explained by the fact that the inhibiting peptide, but not the other, caused dissociation of the complex. Lately, Ward and Becker (1967) showed that the chemotactic factor activates a proesterase in the cell membrane of polymorphonuclear leukocytes.

2. The Low Molecular Weight Factor

Recently, Taylor and Ward (1967) discovered an additional chemotactic factor that is derived from complement. Upon treatment of rabbit serum with plasminogen and streptokinase they observed the appearance of chemotactic activity which was heat labile and associated with a dialyzable substance. Generation of this activity required heat-labile factors in serum and the activity could be derived from a chromatographically obtained serum fraction which contained primarily C′3 and C′5. Following the initial observation, Ward (1967) proceeded to demonstrate that the dialyzable factor was derived from the third component of complement. Treatment of highly purified human C′3 with plasminogen and streptokinase resulted in cleavage of this molecule into two fragments of widely differing size. The smaller fragment, which constitutes the chemotactic factor, has an approximate molecular weight of 6000 and migrates on electrophoresis faster than the intact C′3 molecule. It is distinct from C′3 derived anaphylatoxin in that it lacks the capacity to release histamine from mast cells. Anaphylatoxin, on the other hand, lacks chemotactic activity.

C. COMPLEMENT-DEPENDENT, NONCYTOTOXIC HISTAMINE RELEASE FROM RAT PERITONEAL MAST CELLS

In their studies on histamine release, Austen et al. (1965) uncovered an additional mechanism for the release of histamine from mast cells which is complement dependent. Their in vitro system consists of rat peritoneal mast cells which are sensitized with a rabbit antiserum to rat γ-globulin. The sensitization results from the interaction of rabbit anti-rat γG-globulin with rat γG-globulin on the surface of the mast cells. Histamine release is observed upon addition of fresh serum, whereas heated serum or serum absorbed with an unrelated immune precipitate failed to sustain the reaction. A detailed analysis of the possible requirement of complement disclosed that at least five components participate in this reaction, namely C′1, C′2, C′3, C′4, and C′5 (Austen and Becker,

1966). This conclusion was based upon the restoration of histamine-releasing capacity by the addition of highly purified complement components to sera deficient in one or more of these components. Thus, a reagent deficient in C'1 was active only when C'1a was previously attached to the sensitized mast cells. Similarly, human serum genetically deficient in C'2 and inactive in both the hemolytic and the histamine-releasing systems became reactive with respect to both systems upon reconstitution by purified C'2. Further, a reasonable correspondence between the restoration of histamine-releasing and hemolytic capacity of nephritic sera (which are relatively deficient in C'3) was observed upon addition of highly purified human C'3. Serum treated with 1 M KCNS and thus rendered deficient in C'3, C'4, and C'5 (Dalmasso and Müller-Eberhard, 1966) remained inactive after addition of optimal amounts of C'3 and C'4. Restoration of its histamine-releasing capacity, as of its hemolytic activity, required C'5. Results with genetically C'6-deficient rabbit serum were not consistent; some sera were active (Keller, 1965) but others were not. It is, therefore, not possible at present to decide with certainty whether C'6 and also C'7 are essential factors in this histamine-releasing system.

The complement-dependent mechanism of histamine release is distinct from another mechanism called "homocytotropic-antibody-mediated release" of histamine from mast cells (Austen and Bloch, 1965). The latter reaction requires a unique rat immunoglobulin which differs from γG-, γM-, and γA-globulin and which functions independent of complement (Austen et al., 1965). Despite the difference in mechanism, both the complement-dependent and complement-independent reactions are similarly susceptible to inhibition by phosphonate esters. On the other hand, there is almost complete disparity between the ability of the phosphonates to inhibit complement-dependent histamine release and to inactivate C'1a (Becker and Austen, 1966). One possible explanation considered by Austen and Becker (1966) is that the complement-dependent and complement-independent reactions lead to activation of the same esterase in the membrane of the mast cells. An obvious question arising is whether anaphylatoxin is the actual mediator of the complement-dependent reaction on mast cells. If anaphylatoxin can be produced from rat C'5 as it can from guinea pig and human C'5, then it may well be the active agent in the reaction described by Austen and Becker (1966). Should C'6 and C'7 also be needed in the complement-dependent mechanism, then the actual mediator may well be the C'5,6,7 complex, which was found by Ward and Becker (1967) to activate a proesterase in the membrane of polymorphonuclear leukocytes.

It is quite conceivable that the chemical mode of action of C'5,6,7 is independent of cell type and that the biological effects of this complex depend on the nature of the target cells.

D. IMMUNE ADHERENCE AND OPSONIZATION

The phenomenon of immune adherence was described by R. A. Nelson (1953) and has been reviewed extensively by D. S. Nelson (1963). It is based on the ability of antigen–antibody–complement complexes to adhere to the surface of nonsensitized particles such as red cells, leukocytes, platelets, and starch granules. R. A. Nelson (1953, 1956) and Robineaux and R. A. Nelson (1955) demonstrated that the adherence phenomenon can be observed in vivo and linked it to the mechanism of phagocytosis. Sensitized bacteria or viruses that become attached in the circulation to erythrocytes are an easier prey for phagocytes than unattached particles. Furthermore, it seems probable that opsonization by complement is based on immune adherence: A particle that tends to adhere to the surface of a phagocyte may become engulfed more readily. In studying the requirements for the immune adherence phenomenon, Nishioka and Linscott (1963) found that the critical component was cell-bound C'3; cells in the state EAC'1a,4,2a,3, EAC'1a,4,3, and EAC'4,3 were equally reactive. The chemical mode of the reaction is largely unknown, although it has been speculated that it is enzymatic in nature (D. S. Nelson, 1965). In support of this view, Basch (1965) found that the reaction is inhibitable by simple peptides containing an aromatic amino acid. Similar compounds have been shown to be hydrolyzed by C'3-containing cells, and this finding has pointed out the existence of the C'3-dependent peptidase (Cooper and Becker, 1967). Whether or not this enzyme is responsible for immune adherence is a matter of future exploration.

Immune adherence has been utilized in the past as a convenient and sensitive method for the quantitation of C'3. Using radioactively labeled C'3 it was possible to evaluate the sensitivity of this method. In earlier experiments more than 150 molecules of bound C'3 per cell were required for a weakly positive immune adherence reaction (Müller-Eberhard et al., 1966a). Lately, using a more sensitive technique, it was found that at least 60–100 bound molecules per cell are needed and that the intensity of the reaction reaches its maximum at approximately 1000 bound C'3 molecules per cell (Cooper and Müller-Eberhard, 1967b). Any larger number of molecules cannot be detected by this method, which is a serious limitation since several hundred thousand C'3 molecules may become bound by a single cell.

Promotion of phagocytosis by complement *in vitro* was studied by R. A. Nelson (1962) who found that opsonization of red cells was achieved by cell-bound C′1,4,2,3; C′3 being the critical component. In many instances, heavy coating of particles by specific antibody is sufficient for opsonization. Larger particles, however, seem to require, in addition, cell-bound C′3. There are, of course, exceptions as, for instance, in the case of red cells sensitized with antibodies to the D antigen which do not bind complement. These cells can be engulfed by monocytes provided that they are brought into direct contact with the phagocytic cell by some method such as centrifugation (Archer, 1965). Participation of complement in phagocytosis *in vivo* was demonstrated by a striking reduction in the clearance rate of antibody-coated *Escherichia coli* and sensitized rat erythrocytes in mice decomplemented with aggregated γ-globulin. In addition, red cells sensitized with the F(ab)₂ fragment of pepsin-digested antibody were eliminated from the circulation of mice at a rate characteristic for untreated red cells (Spiegelberg *et al.*, 1963). The complement-fixing ability of this fragment is known to be limited.

E. IMMUNOCONGLUTINATION AND CONGLUTINATION

Immunoconglutinin is an antibody to bound complement. It arises in response to immunization and in the course of various diseases, including bacterial infections (Coombs *et al.*, 1961). It is directed toward antigenic determinants which are hidden in the native complement components and which become accessible upon their fixation to antigen–antibody complexes. Immunoconglutinin directed to autologous complement constitutes the prototype of an autoantibody. It may be obtained as an iso- or heterospecific antibody by injecting particles coated with complement from another individual of the same species or from a different species. Whereas iso- and heterostimulated immunoconglutinin is first of γM- and then of γG-antibody type, autostimulated immunoconglutinin is invariably of the γM type and remains so even after repeated stimulation (Lachmann and Coombs, 1965). Lachmann (1966) showed that immunoconglutinins are specifically directed either to bound C′3 or to bound C′4. Some may be directed to that portion of the C′4 molecule that becomes inaccessible after fixation of C′2. The biological significance of the immunoconglutinins is not entirely clear; according to Coombs *et al.* (1961), they enhance host resistance to infection.

Conglutinin is a unique protein occurring in bovine serum, resembling in activity immunoconglutinin but differing immunochemically from it in that it is not related to γ-globulin. Lachmann (1962) isolated the protein, and Lachmann and Richards (1964) determined some of its molecular parameters. Conglutinin is a highly asymmetrical molecule with a molecular weight of 746,000, a sedimentation coefficient, $s^0_{20,w}$, of 7.8 S and a diffusion coefficient of 0.9×10^{-7} cm.^{2}/second. By electron microscopy the molecular dimensions were found to be 450×60 Å. Chemically, the most remarkable feature was a glycine content of 18% (mole/mole) (Lachmann and Coombs, 1965). Conglutinin is capable of attaching itself to the carbohydrate moiety of bound C'3 and to zymosan in the presence of calcium ions. Conglutination, i.e., agglutination of complement-coated particles by conglutinin, is based on this reaction. Leon et al. (1966) investigated the chemical specificity of conglutinin by studying the inhibition of conglutination by extracts from zymosan and other strains of yeast. The active product was identified as a mannan–peptide complex, digestion of which by pronase did not affect its inhibitory activity. Mannans from 4 of 60 strains of yeast tested were found to be inhibitory. The grouping in the mannans responsible for their reactivity was an α-1,2-mannotriose. N-Acetyl-D-glucosamine also was found to be a potent inhibitor. It is probable that similar carbohydrate groups are responsible for the conglutinogen character of bound C'3.

Lately, a new factor was detected which plays an essential role in the conglutinin system (Lachmann and Müller-Eberhard, 1968). Its requirement became apparent through the negative reaction of conglutinin with EAC'1a,4,2a,3 cells which were built up with purified human complement components. Previously this intermediate was found to be strongly reactive when prepared with appropriate serum reagents (Lachmann and Liske, 1966). Nonreactive, built-up EAC'1a,4,2a,3 could be rendered reactive by treatment with KCNS-inactivated serum which lacks, in addition to C'3 and C'4, also C'5. The responsible factor was shown to be distinct from known complement components and, as it apparently revealed the conglutinogen in bound C'3, it was tentatively called "conglutinogen-activating factor" or "KAF." The new factor was partially purified from human serum and characterized as a β-globulin with a molecular weight of approximately 100,000. Conglutinogen-activating factor is heat stable at 56°C. and resistant to treatment with hydrazine, DFP, soybean trypsin inhibitor, and to a number of other agents known to affect complement components and proteolytic enzymes. It was inactivated by treatment with potassium metaperiodate which suggests the presence of a functionally essential carbohydrate moiety. The action of KAF on bound C'3 was temperature dependent, being optimal at

37°C. This fact as well as the finding that it was not consumed in the course of the reaction suggested that it might be an enzyme and that the conglutinogen is revealed by an enzymatic modification of bound C′3. After treatment with KAF, cell-bound C′3 was much more susceptible to the action of trypsin. Conglutinogen-activating factor action did not appear to lead to fragmentation of bound C′3; however, as shown with radioactively labeled C′3, it weakened the bond between C′3 and cell surface receptors. Invariably KAF treatment led to release of a certain proportion of bound C′3, the molecular weight of which was roughly comparable to that of native C′3. The biological significance of KAF is obscure. Since human serum does not contain conglutinin, its ability to expose the conglutinogen must be accidental.

F.   Lysis of Whole-Blood Clots in Vitro

Although little is known about the mechanism by which blood clots normally lyse, it has been assumed that the enzyme plasmin plays a key role. One of the central questions is how the plasmin system is activated in the intact blood clot. In the course of studies aiming at the elucidation of clot lysis in vitro, observations were made that appear to involve complement in the process (Taylor and Müller-Eberhard, 1968). Although these studies are rudimentary, they are cited here briefly in order to draw attention to the possibility that the complement system or parts thereof may be operative in reactions other than those identified strictly with immune mechanisms.

Blood clots formed from whole blood diluted 1:10 in phosphate buffer in the presence of thrombin usually lyse and shed their red cells within 4 hours at 37°C. Clot lysis was found to require the presence of intact platelets at least in its initial phase. It can be inhibited by a specific antiserum to plasminogen and this inhibition can be overcome by purified plasminogen. In controlling this experiment with specific antisera to other serum proteins, clear-cut inhibition was observed with anti-C′3, anti-C′4, and anti-γM-globulin which could be reversed by addition of the respective purified antigens, i.e., C′3, C′4, and γM-globulin. No such inhibition was noticed with specific antisera to human albumin, α_2-macroglobulin, transferrin, and γG-globulin. Additional observations shed some light on the possible mode of involvement of complement components in this reaction. Human platelets freshly isolated and thoroughly washed could be agglutinated by specific antisera to C′3 and C′4 and to some extent also by an antiserum to γM-globulin. Again, the phenomenon was specific in the sense that it could be inhibited by the respective purified antigens. The other antisera cited above gave negative reactions.

Since C'3 was found among the proteins apparently adhering to the platelets, they were tested for immune adherence and found to be weakly but definitely positive. It is too early to formulate a hypothesis to picture the underlying mechanism. Also, it is completely unknown whether this *in vitro* model has any relevance to *in vivo* situations.

XI. Inherited Complement Deficiencies

A. GUINEA PIG

Early this century a strain of guinea pigs was described that differed from the normal in that complement was missing in their serum. The stock came originally from the Vermont Agricultural Station where it had arisen as a mutation. These animals were studied primarily by Hyde (1932) who found that the heat-stable component of complement was missing which is inactivated by incubation of normal guinea pig serum with zymosan. The defect thus involved the group of the six late acting components, which was then regarded as the third component. It is impossible today to identify the deficient factor, although in view of its thermostability, it might have been either C'3, C'6, or C'7. For some time the discussion centered on the question of whether the deficient animals were more susceptible to infections. Rich (1923) claimed that they were less resistant than normal guinea pigs to experimental infection with *Bacillus cholerae suis*. He also asserted that the deficient animals were unusually susceptible to changes in temperature. In contrast, Hyde (1932) concluded that the deficient animals did just as well as complement-sufficient animals when kept under ideal conditions. However, under adverse environmental conditions, the complement-deficient guinea pigs were the first to succumb. He also concluded, on the basis of his experiments, that complement does not play the role in resistance to infections that test tube experiments had led scientists to believe. Unfortunately, the complement-deficient strain was lost as time elapsed so that it is not possible to re-evaluate these early findings nor to determine the exact nature of the defect in the light of present knowledge.

B. MOUSE

Rosenberg and Tachibana (1962) demonstrated that certain strains of inbred mice were completely devoid of hemolytic complement activity. Genetic analysis indicated that a single gene difference was responsible for the observed variation in complement. The gene locus was designated Hc and the alleles determining the presence and absence of hemolytic complement activity hc' and hc°, respectively (Herzenberg *et al.*, 1963;

Tachibana *et al.*, 1963). Immunization of complement-deficient strains with serum from strains with hemolytic activity led to the formation of antibodies against the gene product, designated hc′ (Erickson *et al.*, 1964). Similar observations were made by Cinader *et al.* (1964) who designated the gene product MuB1. Ninety-nine strains of inbred mice were tested for the presence of MuB1, and the antigen was detected in 44 strains, that is in 61%. These authors also found that the quantity of the antigen MuB1 is always greater in the serum of male than in the serum of female mice. The antigen was characterized as a heat-labile, hydrazine-resistant β-globulin of an approximate molecular weight of 150,000. Evidence was cited indicating that the deficiency was due to lack of synthesis of the respective serum protein and not to synthesis of a chemically modified molecule.

An immunochemical analysis of mouse MuB1 disclosed an antigenic relationship with human C′5 and prompted the attempt to reconstitute the hemolytic activity of complement-deficient mouse plasma by addition of purified human C′5 (Nilsson and Müller-Eberhard, 1967a). Reconstitution could readily be demonstrated. If MuB1 represented the fifth component of mouse complement, the hemolytic activity of human serum should be specifically inhibited by mouse anti-MuB1 and that of mouse serum by anti-human C′5. In both cases, inhibition should be reversed by addition of purified human C′5. Both of these postulates were experimentally verified, and it was, therefore, concluded that mice with an inherited complement deficiency lacked the fifth component of complement.

Evaluation of the passive cutaneous reaction suggested that complement-deficient mice gave a weaker response than normal mice (Ben-Efraim and Cinader, 1964). The resistance to endogenous and exogenous infection with a common mouse pathogen was investigated by Caren and Rosenberg (1966a). When challenged with *Corynebacterium kutscheri*, mice possessing hemolytic complement fared slightly better than complement-deficient mice. Administration of hydrocortisone to mice latently infected with this microorganism led to activation of pseudotuberculosis with equal frequency in mice of both complement types. According to these workers their investigation suggests that at least in one situation the presence of the complete complement system may be advantageous to the mouse. On the other hand, they presented evidence showing that, under normal laboratory conditions, complement-deficient mice have a survival advantage over complement-positive mice during the first 3 weeks of life. They, therefore, concluded that mouse hemolytic complement has a balanced survival value.

C. RABBIT

In 1961, Rother and Rother discovered in Freiburg, Germany, a strain of rabbits which was characterized by an inherited complement defect (Rother and Rother, 1961). They were able to rule out that a powerful inhibitor was responsible for lack of hemolytic activity and defined the defect as involving one of the five last acting components (Volk et al., 1964). With purified human components, reconstitution experiments were carried out and clearly indicated that the defect involved the sixth component of complement (Rother et al., 1966a). Functionally pure C'6 prepared from normal rabbit serum also was able to overcome the deficiency. In addition, it was possible to raise inhibiting antibody to normal C'6 in C'6-deficient animals, thus demonstrating that the defect was caused by lack of synthesis of C'6 or by the synthesis of an immunochemically modified, hemolytically inactive analog.

The deficiency in C'6 explains most of the abnormal immunological reactions that were encountered in these animals, or with their serum. The deficient serum lacks the capacity to kill gram-negative bacteria even after their sensitization with antibody (Rother et al., 1964a), since this requires all the components of complement. It also lacks the ability to generate leukocyte chemotactic activity which is known to depend on the activated C'5,6,7 complex (Ward et al., 1965). In vivo, an impairment of the passive Arthus reactivity was noticed which is most probably related to the strikingly reduced chemotactic activity of their serum (Rother et al., 1964b). By contrast, serological reactions known to depend only on the first four complement components were found to be unimpaired. These are immune adherence, erythrophagocytosis by polymorphonuclear leukocytes in vitro, and enhanced immune clearance of Salmonella typhi in vivo (Rother and Rother, 1965).

Certain other phenomena are presently not sufficiently understood to relate them clearly to the C'6 defect. Unlike their normal littermates, C'6-deficient animals develop markedly suppressed delayed hypersensitivity to tuberculin (Rother et al., 1967a) and with a certain frequency fail to reject skin homografts (Volk et al., 1964; Rother et al., 1967b). These authors feel, therefore, that the complement system participates in some unknown fashion in the immune mechanism of homograft rejection. An unexpected finding was that the C'6-deficient animals following injection with antirabbit kidney antibody produced nephrotoxic nephritis of equal severity as that observed in control animals (Rother et al., 1966b). The amounts of antibody injected were such that the disease was limited to the second phase. It developed also in animals that contained circulating antibody to C'6, thus precluding the possibility that trace amounts of C'6 may have been available for chemotactic attraction of polymorphonuclear leukocytes to the glomeruli (Cochrane et al., 1965). Immunofluorescent studies disclosed in vivo fixation of C'3 in the glomeruli of complement-deficient and normal animals (Rother et al., 1967a). It is conceivable that fixation of C'3 in the glomeruli is sufficient to trigger the events that lead to manifest nephritis.

In 1965, Biro and Garcia (1965) found complement-deficient rabbits in Mexico which later (Biro and Ortega, 1966) were shown to have the same defect as the C'6-deficient Freiburg strain described by Rother and Rother (1961). A limited number of these animals were studied and the following results were obtained. In the active and passive Arthus reaction, hemorrhagic necrosis was absent in the C'6-deficient rabbits, whereas normal controls exhibited a strongly positive reaction. Delayed hypersensitivity to human γ-globulin and skin graft rejection were found to be normal in the deficient animals. The partial disparity between results obtained by the Rothers and those reported by Biro may be due to the comparatively small number of animals used by Biro, which may have been insufficient to permit observation of the abnormal behavior encountered by the Rothers.

D. MAN

1. C'2 Deficiency

A C'2 deficiency state was described in 1960 by Silverstein who called it essential hypocomplementemia. Fourteen similar cases were described in 1964 by Hässig et al. (1964) as a result of a screening study of serum complement levels in more than 41,000 Swiss Army recruits. A hereditary pattern could not be established in any of the respective families studied. The first evidence of an inherited C'2 deficiency was reported by Klemperer et al. in 1966. Using Austen and Beer's C'2 assay (1964) they found a selective decrease in serum C'2 activity in several members from three generations of a kindred. Three individuals had less than three units of C'2 per milliliter serum (normal range, 350–370) and were considered homozygous for the C'2 deficiency and 15 individuals had subnormal levels, ranging between 184 and 316 units/ml.; these were regarded as heterozygous individuals. The defect appeared to be transmitted as an autosomal recessive characteristic. The immune adherence and bactericidal activity of serum was reduced to 10 to 15% of the normal value in the homozygous and to 50% in the heterozygous individuals. Evidence was obtained indicating that the defect was caused by insufficient synthesis of C'2.

A second kindred was discovered when Silverstein's original case was further explored. In this family the propositus was the only homozygous individual (Klemperer *et al.*, 1967). Although his C'2 hemolytic activity was very low, the immune adherence activity of his serum was normal and its bactericidal activity when tested with unsensitized bacteria was near normal. Pickering *et al.* (1966) concluded from their studies that the amount of C'2 present in the serum of C'2-deficient individuals is sufficient to support immune adherence and bactericidal functions *in vitro* and to maintain *in vivo* complement-dependent functions important to host defense mechanisms.

2. Deficiency of C'1 Esterase Inhibitor

Angioneurotic edema which may be inherited as an autosomal dominant trait is accompanied by a biochemical abnormality involving the serum inhibitor of C'1 esterase (Donaldson and Evans, 1963; Landerman *et al.*, 1962) which was previously purified from normal human serum by Pensky *et al.* (1961). The inhibitor is an α_2-globulin having a sedimentation constant of 3 S and a molecular weight of 90,000. It is heat labile and blocks the esterolytic activity of C'1 esterase in a stoichiometric fashion (Lepow, 1965). Two genetic variations of the deficiency have been distinguished: in one variant the inhibitor is totally absent from serum and in the other it is present as an inactive analog which may be detected immunochemically (Rosen *et al.*, 1965). Affected individuals characteristically have active C'1 esterase in their serum and plasma, and the activity is markedly increased during acute attacks. In intervals between attacks, C'2 and C'4 were diminished and during attacks these activities were absent, while C'3 was unchanged (Donaldson and Rosen, 1964; Austen and Sheffer, 1965; Siboo and Laurell, 1965; Laurell *et al.*, 1966). The possibility has been considered that C'1 esterase may serve to increase vascular permeability in this disease and may thus be responsible, at least in part, for the development of the characteristic, and at times hazardous, edema that accompany attacks.

XII. Experimentally Induced Complement Deficiencies

A. Effect of Sex Hormones and Castration

Cinader *et al.* (1964) noticed that the serum of normal female mice contained only half the amount of C'5 present in the serum of male mice. Measuring whole complement levels, Terry *et al.* (1964) found consistently higher levels in male than in female mice. Lately, Caren and Rosenberg (1966b) and Weintraub *et al.* (1966) described the successful manipulation of serum complement activity in mice by administration of sex hormones. Caren and Rosenberg (1966b) measuring whole complement found that male sex hormone increased the normally low levels of complement of females to amounts characteristic for males and that the synthetic feminizing hormone, diethylstilbesterol, lowered the levels of complement in male mice. Males and females were unaffected by their own sex hormones. Weintraub *et al.* (1966) defined this effect as involving one or more of the late acting components of complement. They found that the titer of these late acting components in the serum of male mice, which is normally tenfold higher than that of female mice, fell after castration to the level of female mice. Conversely, the titer of serum from female mice rose drastically after castration; and administration of testerone or estrogen showed a similar effect on the activity of late acting components as reported by Caren and Rosenberg (1966b) for whole complement.

In the course of an analysis of the serological defect in complement-deficient mice, it was found that in normal mice $C'5$ hemolytic activity is appreciably lower in females than in males. However, addition of purified $C'5$ to serum of female origin did not raise the activity to the higher level of serum from male mice unless $C'6$ and $C'7$ were added also. Since complement analysis revealed low hemolytic activity for $C'6$ and $C'7$ in normal female mouse serum, it was concluded that the lower total hemolytic activity of these sera is due to a relative deficiency in $C'5$, $C'6$, and $C'7$ (Nilsson and Müller-Eberhard, 1967a). It appears probable that the effects observed following administration of sex hormones involved primarily these three components.

B. Effect of Cobra Factor

In 1903, Flexner and Noguchi inactivated serum complement by treatment with cobra venom. Later studies showed the effect to be limited to one of the late acting components, and in recent years Klein and Wellensiek (1965b) demonstrated that, in guinea pig serum, $C'3$ was primarily affected. Vogt and Schmidt (1964) working on the production of anaphylatoxin by cobra venom partially purified the responsible venom factor and injected it into rats. They found depletion in the animals of anaphylatoxinogen. Recently, Nelson (1966) purified presumably the same factor from cobra venom and following administration to guinea pigs and rabbits he principally confirmed Klein and Wellensiek's *in vitro* studies by demonstrating *in vivo* depletion of circulating $C'3$.

Exploring the mechanism of C′3 inactivation by cobra factor, the factor was obtained in highly purified form from crude venom of the *Naja naja* (Müller-Eberhard *et al.*, 1966c; Müller-Eberhard, 1967b). It is a carbohydrate (13%)-containing protein with an electrophoretic mobility similar to that of serum β-globulin and an approximate molecular weight of 140,000. The purified cobra factor, although it inactivated C′3 in whole serum, was found to have no effect on purified C′3. Inactivation of C′3 required participation of an as yet unidentified β-globulin of serum which has an *s* rate of approximately 5 S. Evidence was obtained that the C′3 inactivating activity resides in a protein–protein complex formed by the serum factor and the cobra factor, formation of which was inhibited by EDTA. Action of the C′3 inactivator complex on C′3 resulted in fragmentation of the molecule.

Administration of purified cobra factor to rabbits, guinea pigs, and rats resulted in prolonged depletion of circulating C′3. In addition to C′3 activity, the C′3 protein normally detectable in serum by immunochemical methods disappeared from the circulation (Cochrane and Müller-Eberhard, 1967b). The *in vivo* half-life of radioactively labeled cobra factor in rabbits was 32 hours. In C′3-depleted animals, the Arthus reaction was inhibited and accumulation of polymorphonuclear leukocytes in affected vessel walls was decreased. Acute nephrotoxic nephritis was markedly diminished, proteinuria being 0–10% of expected values.

A similar anticomplementary principle was discovered in the venom of the brown recluse spider, *Loxosceles reclusa* (Kniker and Morgan, 1967). The spider venom factor inactivates specifically C′5. Its effect on C′5 appears to be dependent on the presence of at least one serum factor, since purified C′5 was not affected by the venom. Ethylenediaminetetraacetate inhibited the C′5-inactivating principle of the spider venom. The factor has been purified but has not been used for *in vivo* studies because of toxic side effects.

XIII. Conclusion

After a long dormant period, complement research has once again become an active, rapidly moving, and exciting field of the biological sciences. This trend is reflected in the composition of the bibliography cited in this review. Eighty-five per cent of the references listed refer to work published during the past 6 years, 33% thereof representing communications of the past 15 months. From the total body of available information, there is now emerging an understanding, in biochemical terms, of the manner in which the complement components interact to generate the various activities of the complement system. Physical and chemical analyses are revealing greater simplicity on the molecular level than could be predicted a few years ago when the full complexity of the system became apparent. It is likely that the essential features of the molecular basis of complement function will be elucidated within the very next few years. Further advancement will come from a thorough exploration of the biology of complement. New and important insights have already been gained on this subject by both basic and clinical investigators which could only be discussed briefly in this review. New methods and tools are now becoming available to the biologist and clinician for the investigation of the physiogenic and pathogenic role of complement, so that a large body of additional data will soon be at hand. A future reviewer of complement research will undoubtedly be in a better position in presenting the biological aspect.

REFERENCES

Adinolfi, M., Polley, M. J., Hunter, D. A., and Mollison, P. L. (1962). *Immunology* 5, 566.

Alberty, R. A., and Baldwin, R. L. (1951). *J. Immunol.* 66, 725.

Alper, C. A., Levin, A. S., and Rosen, F. S. (1966). *Science* 153, 180.

Amano, T., Inai, S., Seki, Y., Kashiba, S., Fujikawa, K., and Nishimura, S. (1954). *Med. J. Osaka Univ.* 4, 401.

Amiraian, K., and Leikhim, E. J. (1961). *Proc. Soc. Exptl. Biol. Med.* 108, 454.

Archer, G. T. (1965). *Vox Sanguinis* 10, 590.

Austen, K. F., and Becker, E. L. (1966). *J. Exptl. Med.* 124, 397.

Austen, K. F., and Beer, F. (1964). *J. Immunol.* 92, 946.

Austen, K. F., and Bloch, K. J. (1965). *Ciba Found. Symp. Complement* p. 281.

Austen, K. F., and Sheffer, A. L. (1965). *New Engl. J. Med.* 272, 649.

Austen, K. F., Bloch, K. J., Baker, A. R., and Arnason, B. G. (1965). *Proc. Soc. Exptl. Biol. Med.* 120, 542.

Ballieux, R. E., Bernier, G. M., Tominaga, K., and Putnam, F. W. (1964). *Science* 145, 168.

Barbaro, J. F. (1963). *Nature* 199, 819.

Basch, R. S. (1965). *J. Immunol.* 94, 629.

Becker, E. L. (1955). *Nature* 176, 1073.

Becker, E. L. (1956a). *J. Immunol.* 77, 469.

Becker, E. L. (1956b). *J. Immunol.* 77, 462.

Becker, E. L. (1959). *J. Immunol.* 82, 43.

Becker, E. L. (1960). *J. Immunol.* 84, 299.

Becker, E. L., and Austen, K. F. (1966). *J. Exptl. Med.* 124, 379.

Becker, E. L., and Wirtz, G. H. (1959). *Biochim. Biophys. Acta* 35, 291.

Ben-Efraim, S., and Cinader, B. (1964). *J. Exptl. Med.* 120, 925.

Benacerraf, B., Ovary, Z., Bloch, K. J., and Franklin, E. C. (1963). *J. Exptl. Med.* 117, 937.

Bernfeld, P., Bernfeld, M. C., Nisselbaum, J. S., and Fishman, W. H. (1954). *J. Am. Chem. Soc.* 76, 4872.

Bernfeld, P., Tuttle, L. P., and Hubbard, R. W. (1961). *Arch. Biochem. Biophys.* 92, 232.

Biro, C. E., and Garcia, G. (1965). *Immunology* **8**, 411.
Biro, C. E., and Ortega, M. L. (1966). *Arch. Inst. Cardiol. Mex.* **36**, 166.
Bladen, H. A., Evans, R. T., and Mergenhagen, S. E. (1966). *J. Bacteriol.* **91**, 2377.
Bokisch, V., and Müller-Eberhard, H. J. (1968). To be published.
Bokisch, V., Müller-Eberhard, H. J., and Cochrane, C. G. (1968). To be published.
Bordet, J., and Gengou, O. (1901). *Ann. Inst. Pasteur* **15**, 289.
Borsos, T., and Rapp, H. J. (1965a). *J. Immunol.* **94**, 510.
Borsos, T., and Rapp, H. J. (1965b). *Science* **150**, 505.
Borsos, T., and Rapp, H. J. (1965c). *J. Immunol.* **95**, 559.
Borsos, T., and Rapp, H. J. (1966). *Immunochemistry* **3**, 496 (abstr.).
Borsos, T., Rapp, H. J., and Mayer, M. M. (1961a). *J. Immunol.* **87**, 326.
Borsos, T., Rapp, H. J., and Mayer, M. M. (1961b). *J. Immunol.* **87**, 310.
Borsos, T., Dourmashkin, R. R., and Humphrey, J. H. (1964a). *Nature* **202**, 251.
Borsos, T., Rapp, H. J., and Walz, U. L. (1964b). *J. Immunol.* **92**, 108.
Borsos, T., Rapp, H. J., and Crisler, C. (1965). *J. Immunol.* **94**, 662.
Boursnell, J. C., Coombs, R. R. A., and Rizk, V. (1953). *Biochem. J.* **55**, 745.
Boyden, S. (1962). *J. Exptl. Med.* **115**, 453.
Boyer, J. T. (1964). *Proc. 10th Congr. Intern. Soc. Hematol., Stockholm* (abstr.).
Buchner, H. (1889). *Zentr. Bakteriol.* **5**, 817.
Budzko, D. B., and Müller-Eberhard, H. J. (1968). To be published.
Calcott, M. A., and Müller-Eberhard, H. J. (1968). To be published.
Caren, L. D., and Rosenberg, L. T. (1966a). *J. Exptl. Med.* **124**, 689.
Caren, L. D., and Rosenberg, L. T. (1966b). *Science* **152**, 782.
Carpenter, C. B., and Gill, T. J. (1966). *Immunology* **10**, 355.
Cebra, J. J. (1963). *In* "Conceptual Advances in Immunology and Oncology" (R. W. Cumley, D. M. Aldridge, J. Haroz, and J. McCay, eds.), p. 220. Harper & Row, New York.
Chan, P. C. Y., and Cebra, J. J. (1966). *Immunochemistry* **3**, 496 (abstr.).
Cinader, B., Dubiski, S., and Wardlaw, A. C. (1964). *J. Exptl. Med.* **120**, 897.
Cochrane, C. G., and Müller-Eberhard, H. J. (1967a). *J. Exptl. Med.* In press.
Cochrane, C. G., and Müller-Eberhard, H. J. (1967b). *Federation Proc.* **26**, 362.
Cochrane, C. G., Unanue, E. R., and Dixon, F. J. (1965). *J. Exptl. Med.* **122**, 99.
Colten, H. R., Borsos, T., and Rapp, H. J. (1966). *Proc. Natl. Acad. Sci. U.S.* **56**, 1158.
Coombs, R. R. A., Coombs, A. M., and Ingram, D. G. (1961). "The Serology of Conglutination and its Relation to Disease." Blackwell, Oxford.
Cooper, N. R. (1967). *J. Immunol.* **98**, 132.
Cooper, N. R., and Becker, E. L. (1967). *J. Immunol.* **98**, 119.
Cooper, N. R., and Müller-Eberhard, H. J. (1967a). *Federation Proc.* **26**, 361 (abstr.).
Cooper, N. R., and Müller-Eberhard, H. J. (1967b). *Immunochemistry.* In press.
Creighton, T. E., and Yanofsky, C. (1966). *J. Biol. Chem.* **241**, 980.
Cushman, W. F., Becker, E. L., and Wirtz, G. H. (1957). *J. Immunol.* **79**, 80.
Dalmasso, A. P., and Müller-Eberhard, H. J. (1964). *Proc. Soc. Exptl. Biol. Med.* **117**, 643.
Dalmasso, A. P., and Müller-Eberhard, H. J. (1966). *J. Immunol.* **97**, 680.
Dalmasso, A. P., and Müller-Eberhard, H. J. (1967). *Immunology* **13**, 293.
Davies, G. E. (1963). *Immunology* **6**, 561.
Dias Da Silva, W., and Lepow, I. H. (1966a). *Immunochemistry* **3**, 497 (abstr.).
Dias Da Silva, W., and Lepow, I. H. (1966b). *J. Immunol.* **95**, 1080.

Dias Da Silva, W., and Lepow, I. H. (1967). *J. Exptl. Med.* **125**, 921.
Donaldson, V. H., and Evans, R. R. (1963). *Am. J. Med.* **35**, 37.
Donaldson, V. H., and Rosen, F. S. (1964). *J. Clin. Invest.* **43**, 2204.
Erickson, R. P., Tachibana, D. K., Herzenberg, L. A., and Rosenberg, L. T. (1964). *J. Immunol.* **92**, 611.
Fischer, H. (1965). *Bull. Schweiz. Akad. Med. Wiss.* **21**, 471.
Fischer, H., and Haupt, I. (1961). *Z. Naturforsch.* **16b**, 321.
Fischer, H., and Haupt, I. (1965). *In* "Immunochemie" (O. Westphal and L. T. Haak, eds.), p. 284. Springer, Berlin.
Fjellström, K. E., and Müller-Eberhard, H. J. (1968). To be published.
Flexner, S., and Noguchi, H. (1903). *J. Exptl. Med.* **6**, 277.
Frank, M. M., Rapp, H. J., and Borsos, T. (1964). *J. Immunol.* **93**, 409.
Friedberger, E. (1910). *Z. Immunitatesforsch.* **4**, 636.
Friedberger, E., Mita, S., and Kumagai, T. (1913). *Z. Immunitatesforsch.* **17**, 506.
Goldberg, B., and Green, H. (1959). *J. Exptl. Med.* **109**, 505.
Good, R. A., Finstad, J., Pollara, B., and Gabrielsen, A. E. (1966). *In* "Phylogeny of Immunity" (R. T. Smith, P. A. Miescher, and R. A. Good, eds.), p. 149. Univ. of Florida Press, Gainesville, Florida.
Green, H., Fleischer, R. A., Barrow, P., and Goldberg, B. (1959a). *J. Exptl. Med.* **109**, 511.
Green, H., Barrow, P., and Goldberg, B. (1959b). *J. Exptl. Med.* **110**, 699.
Grey, H. M., and Kunkel, H. G. (1964). *J. Exptl. Med.* **120**, 253.
Griffin, D., Tachibana, D. K., Nelson, B., and Rosenberg, L. T. (1967). *Immunochemistry* **4**, 23.
Hadding, U., and Müller-Eberhard, H. J. (1968). Submitted to *J. Exptl. Med.*
Hadding, U., Müller-Eberhard, H. J., and Dalmasso, A. P. (1966). *Federation Proc.* **25**, 485.
Hässig, A., Borel, J. F., Ammann, P., Thöni, M., and Bütler, R. (1964). *Pathol. Microbiol.* **27**, 542.
Hahn, F., and Lange, A. (1956). *Deut. Med. Wochschr.* **81**, 1269.
Haines, A. L., and Lepow, I. H. (1964a). *J. Immunol.* **92**, 456.
Haines, A. L., and Lepow, I. H. (1964b). *J. Immunol.* **92**, 468.
Haines, A. L., and Lepow, I. H. (1964c). *J. Immunol.* **92**, 479.
Harboe, M. (1964). *Brit. J. Haematol.* **10**, 339.
Harboe, M., Müller-Eberhard, H. J., Fudenberg, H., Polley, M. J., and Mollison, P. L. (1963). *Immunology* **6**, 412.
Herzenberg, L. A., Tachibana, D. K., and Rosenberg, L. T. (1963). *Genetics* **48**, 711.
Hinz, C. F. (1966). *Progr. Hematol.* **5**, 60.
Hinz, C. F., and Mollner, A. M. (1962). *J. Clin. Invest.* **41**, 1365 (abstr.).
Hinz, C. F., Jordan, W. S., and Pillemer, L. (1956). *J. Clin. Invest.* **35**, 453.
Hoffmann, L. G. (1960). Ph.D. Thesis, Johns Hopkins School of Hygiene and Public Health, Baltimore, Maryland.
Humphrey, J. H., and Dourmashkin, R. R. (1965). *Ciba Found. Symp. Complement* p. 175.
Hyde, R. R. (1932). *Am. J. Hyg.* **15**, 824.
Inoue, K., and Nelson, R. A. (1965). *J. Immunol.* **95**, 355.
Inoue, K., and Nelson, R. A. (1966). *J. Immunol.* **96**, 386.
Ishizaka, K., Ishizaka, T., and Sugahara, T. (1962). *J. Immunol.* **88**, 690.

Ishizaka, T., Ishizaka, K., Borsos, T., and Rapp, H. J. (1966a). *J. Immunol.* **97**, 716.
Ishizaka, T., Ishizaka, K., Salmon, S., and Fudenberg, H. (1966b). *Federation Proc.* **25**, 489 (abstr.).
Jenkins, D. E., Christenson, W. N., and Engle, R. L. (1966). *J. Clin. Invest.* **45**, 796.
Jensen, J. (1967). *Science* **155**, 1122.
Kabat, E. A. (1956). "Blood Group Substances, their Chemistry and Immunochemistry," p. 290. Academic Press, New York.
Keller, R. (1965). *Experientia* **21**, 295.
Klein, P. G. (1965). *In* "Immunochemie" (O. Westphal and L. T. Haak, eds.), p. 330. Springer, Berlin.
Klein, P. G., and Burkholder, P. M. (1960). *J. Exptl. Med.* **111**, 107.
Klein, P. G., and Wellensiek, H. J. (1965a). *Intern. Rev. Exptl. Pathol.* **4**, 245.
Klein, P. G., and Wellensiek, H. J. (1965b). *Immunology* **8**, 590.
Klemperer, M. R., Gotoff, S. P., Alper, C. A., Levin, A. S., and Rosen, F. S. (1965). *Pediatrics* **35**, 765.
Klemperer, M. R., Woodworth, H. C., Rosen, F. S., and Austen, K. F. (1966). *J. Clin. Invest.* **45**, 880.
Klemperer, M. R., Austen, K. F., and Rosen, F. S. (1967). *J. Immunol.* **98**, 72.
Klun, M. J., and Muschel, L. H. (1966). *Nature* **212**, 159.
Kniker, W. T., and Morgan, P. N. (1967). *Federation Proc.* **26**, 362.
Kohler, P., and Müller-Eberhard, H. J. (1967). *Clin. Res.* **15**, 296 (abstr.).
Lachmann, P. J. (1962). *Immunology* **5**, 687.
Lachmann, P. J. (1966). *Immunology* **11**, 263.
Lachmann, P. J., and Coombs, R. R. A. (1965). *Ciba Found. Symp. Complement* p. 242.
Lachmann, P. J., and Liske, R. (1966). *Immunology* **11**, 243.
Lachmann, P. J., and Müller-Eberhard, H. J. (1967). *J. Immunol.* In press.
Lachmann, P. J., and Richards, C. B. (1964). *Immunochemistry* **1**, 37.
Landerman, N. S., Webster, M. E., Becker, E. L., and Ratcliffe, H. E. (1962). *J. Allergy* **33**, 330.
Laurell, A. B., and Siboo, R. (1966). *Acta Pathol. Microbiol. Scand.* **68**, 230.
Laurell, A. B., Lundh, B., Malmquist, J., and Siboo, R. (1966). *Clin. Exptl. Immunol.* **1**, 13.
Leddy, J. P., Hill, R. W., Swisher, S. N., and Vaughan, J. H. (1963). *In* "Immunopathology" (P. Grabar and P. A. Miescher, eds.), Vol. 3, p. 318. Benno Schwabe, Basel.
Leddy, J. P., Bakemeier, R. F., and Vaughan, J. H. (1965). *J. Clin. Invest.* **44**, 1066 (abstr.).
Leon, M. A. (1965). *Science* **147**, 1034.
Leon, M. A., Yokohari, R., and Itoh, C. (1966). *Immunochemistry* **3**, 499 (abstr.).
Lepow, I. H. (1965). *In* "Immunological Diseases" (M. Samter, ed.), p. 188. Little, Brown, Boston, Massachusetts.
Lepow, I. H. (1967). Personal communication.
Lepow, I. H., Ratnoff, O. D., Rosen, F. S., and Pillemer, L. (1956a). *Proc. Soc. Exptl. Biol. Med.* **92**, 32.
Lepow, I. H., Ratnoff, O. D., and Pillemer, L. (1956b). *Proc. Soc. Exptl. Biol. Med.* **92**, 111.
Lepow, I. H., Ratnoff, O. D., and Levy, L. R. (1958). *J. Exptl. Med.* **107**, 451.
Lepow, I. H., Naff, G. B., Todd, E. W., Pensky, J., and Hinz, C. F. (1963). *J. Exptl. Med.* **117**, 983.

Lepow, I. H., Todd, E. W., Smink, R. D., and Pensky, J. (1965a). *Federation Proc.* **24**, 446 (abstr.).

Lepow, I. H., Naff, G. B., and Pensky, J. (1965b). *Ciba Found. Symp. Complement* p. 74.

Leschly, W. (1914). "Studier over Komplement." Stiftsboktrykkeriet, Aarhus, Denmark.

Levine, L. (1955). *Biochim. Biophys. Acta* **18**, 283.

Levine, L., Osler, A. G., and Mayer, M. M. (1953). *J. Immunol.* **71**, 374.

Levine, L., Wasserman, E., and Mills, S. (1961). *J. Immunol.* **86**, 675.

Linscott, W. D., and Cochrane, C. G. (1964). *J. Immunol.* **93**, 972.

Linscott, W. D., and Nishioka, K. (1963). *J. Exptl. Med.* **118**, 795.

Lundh, B. (1964). *Scand. J. Clin. Lab. Invest.* **16**, 108.

McConahey, P. J., and Dixon, F. J. (1966). *Intern. Arch. Allergy Appl. Immunol.* **29**, 185.

Manni, J. A., and Müller-Eberhard, H. J. (1968). To be published.

Mansa, B. (1964). *Acta Pathol. Microbiol. Scand.* **62**, 299.

Mardiney, M. R., and Müller-Eberhard, H. J. (1964). *J. Immunol.* **94**, 877.

Mardiney, M. R., Dalmasso, A., and Müller-Eberhard, H. J. (1968). To be published. *Intern. Arch. Allergy Appl. Immunol.*

Masouredis, S. P. (1960). *J. Clin. Invest.* **39**, 1450.

Mayer, M. M. (1961a). *In* "Immunochemical Approaches to Problems in Microbiology" (M. Heidelberger and O. J. Plescia, eds.), p. 268. Rutgers Univ. Press, New Brunswick, New Jersey.

Mayer, M. M. (1961b). *In* "Experimental Immunochemistry" (E. A. Kabat and M. M. Mayer, eds.), p. 133. Thomas, Springfield, Illinois.

Mayer, M. M. (1965). *Ciba Found. Symp. Complement* p. 4.

Mayer, M. M., and Miller, J. A. (1965). *Immunochemistry* **2**, 71.

Mayer, M. M., Levine, L., Rapp, H. J., and Marucci, A. A. (1954). *J. Immunol.* **73**, 443.

Morse, J. H., and Christian, C. L. (1964). *J. Exptl. Med.* **119**, 195.

Müller-Eberhard, H. J. (1961). *Acta Soc. Med. Upsalien.* **66**, 152.

Müller-Eberhard, H. J. (1962). *In* "Mechanism of Cell and Tissue Damage Produced by Immune Reactions" (P. Grabar and P. A. Miescher, eds.), p. 23. Benno Schwabe, Basel.

Müller-Eberhard, H. J. (1967a). *In* "Methods in Immunology and Immunochemistry" (C. A. Williams and M. W. Chase, eds.). Academic Press, New York. In press.

Müller-Eberhard, H. J. (1967b). *Federation Proc.* **26**, 744 (abstr.).

Müller-Eberhard, H. J., and Biro, C. E. (1963). *J. Exptl. Med.* **118**, 447.

Müller-Eberhard, H. J., and Calcott, M. A. (1966). *Immunochemistry* **3**, 500 (abstr.).

Müller-Eberhard, H. J., and Calcott, M. A. (1968). To be published.

Müller-Eberhard, H. J., and Kunkel, H. G. (1961). *Proc. Soc. Exptl. Biol. Med.* **106**, 291.

Müller-Eberhard, H. J., and Lepow, I. H. (1965). *J. Exptl. Med.* **121**, 819.

Müller-Eberhard, H. J., and Nilsson, U. R. (1960). *J. Exptl. Med.* **111**, 217.

Müller-Eberhard, H. J., Nilsson, U. R., and Aronsson, T. (1960). *J. Exptl. Med.* **111**, 201.

Müller-Eberhard, H. J., Dalmasso, A. P., and Calcott, M. A. (1966a). *J. Exptl. Med.* **123**, 33.

Müller-Eberhard, H. J., Polley, M. J., and Calcott, M. A. (1966b). *Immunochemistry* **3**, 500 (abstr.).

Müller-Eberhard, H. J., Nilsson, U. R., Dalmasso, A. P., Polley, M. J., and Calcott, M. A. (1966c). *Arch. Pathol.* **82**, 205.

Müller-Eberhard, J. H., Polley, M. J., and Nilsson, U. R. (1966d). In "Immunopathology" (P. Grabar and P. A. Miescher, eds.), Vol. 4, p. 421. Benno Schwabe, Basel.

Müller-Eberhard, H. J., Polley, M. J., and Calcott, M. A. (1967a). *J. Exptl. Med.* **125**, 359.

Müller-Eberhard, H. J., Calcott, M. A., and Grey, H. M. (1967b). To be published.

Muschel, L. H. (1965). *Ciba Found. Symp. Complement* p. 155.

Muschel, L. H., and Jackson, J. E. (1966). *J. Immunol.* **97**, 46.

Muschel, L. H., Carey, W. F., and Baron, L. S. (1959). *J. Immunol.* **82**, 38.

Naff, G. B., Pensky, J., and Lepow, I. H. (1964). *J. Exptl. Med.* **119**, 593.

Nelson, D. S. (1963). *Advan. Immunol.* **3**, 131.

Nelson, D. S. (1965). *Ciba Found. Symp. Complement* p. 222.

Nelson, R. A. (1953). *Science* **118**, 733.

Nelson, R. A. (1956). *Proc. Roy. Soc. Med.* **49**, 55.

Nelson, R. A. (1962). In "Mechanism of Cell and Tissue Damage Produced by Immune Reactions" (P. Grabar and P. A. Miescher, eds.), p. 245. Benno Schwabe, Basel.

Nelson, R. A. (1965). In "The Inflammatory Process" (B. W. Zweifach, R. T. McCluskey, and L. H. Grant, eds.), Chapt. 25. Academic Press, New York.

Nelson, R. A. (1966). *Surv. Ophthalmol.* **11**, 498.

Nelson, R. A., Jensen, J., Gigli, I., and Tamura, N. (1966). *Immunochemistry* **3**, 111.

Nilsson, U. R. (1967). *Acta Pathol. Microbiol. Scand.* (in press).

Nilsson, U. R., and Müller-Eberhard, H. J. (1965). *J. Exptl. Med.* **122**, 277.

Nilsson, U. R., and Müller-Eberhard, H. J. (1966). *Immunochemistry* **3**, 500 (abstr.).

Nilsson, U. R., and Müller-Eberhard, H. J. (1967a). *J. Exptl. Med.* **125**, 1.

Nilsson, U. R., and Müller-Eberhard, H. J. (1967b). *Immunology* **13**, 101.

Nishioka, K., and Linscott, W. D. (1963). *J. Exptl. Med.* **118**, 767.

Opferkuch, W., and Klein, P. G. (1964). *Immunology* **7**, 261.

Osawa, E., and Muschel, L. H. (1964). *J. Exptl. Med.* **119**, 41.

Osler, A. G., Randall, H. G., Hill, B. M., and Ovary, Z. (1959). *J. Exptl. Med.* **110**, 311.

Peetoom, F., and Pondman, K. W. (1963). *Vox Sanguinis* **8**, 605.

Peetoom, F., van der Hart, M., and Pondman, K. W. (1964). *Vox Sanguinis* **9**, 85.

Pensky, J., Levy, L. R., and Lepow, I. H. (1961). *J. Biol. Chem.* **236**, 1674.

Pickering, R. J., Muschel, L. H., Mergenhagen, S. E., and Good, R. A. (1966). *Lancet* **ii**, 356.

Plescia, O. J., Amiraian, K., and Cavallo, G. (1957). *Federation Proc.* **16**, 429 (abstr.).

Polley, M. J. (1966). *Federation Proc.* **25**, 485 (abstr.).

Polley, M. J., and Mollison, P. L. (1961). *Transfusion* **1**, 9.

Polley, M. J., and Müller-Eberhard, H. J. (1965a). *Federation Proc.* **24**, 446 (abstr.).

78   HANS J. MÜLLER-EBERHARD

Polley, M. J., and Müller-Eberhard, H. J. (1965b). *Science* **148**, 1728.
Polley, M. J., and Müller-Eberhard, H. J. (1966). *Immunochemistry* **3**, 501 (abstr.).
Polley, M. J., and Müller-Eberhard, H. J. (1967a). To be published.
Polley, M. J., and Müller-Eberhard, H. J. (1967b). *J. Exptl. Med.* **126**, 1013.
Polley, M. J., and Müller-Eberhard, H. J. (1968). *J. Exptl. Med.* In press.
Pondman, K. W., and Peetoom, F. (1964). *Immunochemistry* **1**, 65.
Rapp, H. J. (1958). *Science* **127**, 234.
Rapp, H. J., and Borsos, T. (1963). *J. Immunol.* **91**, 826.
Ratnoff, O. D., and Lepow, I. H. (1957). *J. Exptl. Med.* **106**, 327.
Ratnoff, O. D., and Lepow, I. H. (1963). *J. Exptl. Med.* **118**, 681.
Ratnoff, O. D., and Naff, G. B. (1967). *J. Exptl. Med.* **125**, 337.
Reiss, A. M., and Plescia, O. J. (1963). *Science* **141**, 812.
Rich, F. A. (1923). *Vermont, Univ. Agr. Expt. Sta. Bull.* **230**, 1.
Robineaux, R., and Nelson, R. A. (1955). *Ann. Inst. Pasteur* **89**, 254.
Rochna, E. M., and Hughes-Jones, N. C. (1965). *Vox Sanguinis* **10**, 675.
Rommel, F. A., and Stolfi, R. L. (1966). *Immunochemistry* **3**, 502 (abstr.).
Rosen, F. S., Charache, P., Pensky, J., and Donaldson, V. (1965). *Science* **148**, 957.
Rosenberg, L. T., and Tachibana, D. K. (1962). *J. Immunol.* **89**, 861.
Rosse, W. F., and Dacie, J. V. (1966). *J. Clin. Invest.* **45**, 736.
Rosse, W. F., Dourmashkin, R., and Humphrey, J. H. (1966). *J. Exptl. Med.* **123**, 969.
Rother, K., and Rother, U. (1965). *Proc. Soc. Exptl. Biol. Med.* **119**, 1055.
Rother, K., McCluskey, R. T., and Rother, U. (1967a). *Federation Proc.* **26**, 787.
Rother, U., Ballantyne, D. L., Jr., Cohen, C., and Rother, K. (1967b). *J. Exptl. Med.* **126**, 565.
Rother, K., Rother, U., Petersen, K. F., Gemsa, D., and Mitze, F. (1964a). *J. Immunol.* **93**, 319.
Rother, K., Rother, U., and Schindera, F. (1964b). *Z. Immunitaetsforsch.* **126**, 473.
Rother, K., Rother, U., Müller-Eberhard, H. J., and Nilsson, U. R. (1966a). *J. Exptl. Med.* **124**, 773.
Rother, K., Vassalli, P., Rother, U., and McCluskey, R. T. (1966b). *Federation Proc.* **25**, 309.
Rother, U., and Rother, K. (1961). *Z. Immunitaetsforsch.* **121**, 224.
Schultze, H. E., Heide, K., and Haupt, H. (1962). *Klin. Wochschr.* **40**, 729.
Schur, P. H., and Becker, E. L. (1963). *J. Exptl. Med.* **118**, 891.
Schur, P. H., and Christian, G. D. (1964). *J. Exptl. Med.* **120**, 531.
Sears, D. A., Weed, R. I., and Swisher, S. N. (1964). *J. Clin. Invest.* **43**, 975.
Shin, H. S., and Mayer, M. M. (1967). *Fed. Proc.* **26**, 361.
Siboo, R., and Laurell, A. B. (1965). *Acta Pathol. Microbiol. Scand.* **65**, 413.
Silverstein, A. M. (1960). *Blood* **16**, 1338.
Sitomer, G., Stroud, R. M., and Mayer, M. M. (1966). *Immunochemistry* **3**, 57.
Spiegelberg, H. L., Miescher, P. A., and Benacerraf, B. (1963). *J. Immunol.* **90**, 751.
Stecher, V. J., Jacobson, E. B., and Thorbecke, G. J. (1965). *Federation Proc.* **24**, 447 (abstr.).
Stecher, V. J., Morse, J. H., and Thorbecke, G. J. (1967). *Proc. Soc. Exptl. Biol. Med.* **124**, 433.
Stegemann, H., Vogt, W., and Friedberg, K. D. (1964a). *Z. Physiol. Chem.* **337**, 269.
Stegemann, H., Bernhard, G., and O'Neil, J. A. (1964b). *Z. Physiol. Chem.* **339**, 9.
Stegemann, H., Hillebrecht, R., and Rien, W. (1965). *Z. Physiol. Chem.* **340**, 11.
Steinbuch, V. J., Quentin, M., and Pejaudier, L. (1963). *Nature* **200**, 262.

Stelos, P., and Talmage, D. W. (1957). *J. Infect. Diseases* **100**, 126.

Stolfi, R. (1967). *Federation Proc.* **26**, 362.

Stratton, F. (1961). *Nature* **190**, 240.

Stroud, R. M., Austen, K. F., and Mayer, M. M. (1965). *Immunochemistry* **2**, 219.

Stroud, R. M., Mayer, M. M., Miller, J. A., and McKenzie, A. T. (1966). *Immunochemistry* **3**, 163.

Tachibana, D. K., Ulrich, M., and Rosenberg, L. T. (1963). *J. Immunol.* **91**, 230.

Talmage, D. W., Freter, G. G., and Taliaferro, W. H. (1956). *J. Infect. Diseases* **98**, 300.

Taranta, A., and Franklin, E. C. (1961). *Science* **134**, 1981.

Taranta, A., Weiss, H. S., and Franklin, E. C. (1961). *Nature* **189**, 239.

Taylor, A. B., and Leon, M. A. (1959). *J. Immunol.* **83**, 284.

Taylor, F. B., and Fudenberg, H. (1964). *Immunology* **7**, 319.

Taylor, F. B., and Müller-Eberhard, H. J. (1968). *Nature.* In press.

Taylor, F. B., and Ward, P. A. (1967). *J. Exptl. Med.* **126**, 149.

Terry, W. D., and Fahey, J. L. (1964). *Science* **146**, 400.

Terry, W. D., Borsos, T., and Rapp, H. J. (1964). *J. Immunol.* **92**, 576.

Thorbecke, G. J., Hochwald, G. M., van Furth, R., Müller-Eberhard, H. J., and Jacobson, E. B. (1965). *Ciba Found. Symp. Complement* p. 99.

Tiselius, A., Hjertén, S., and Levin, Ö. (1956). *Arch. Biochem. Biophys.* **65**, 132.

Vogt, W., and Schmidt, G. (1964). *Experientia* **20**, 207.

Volk, H., Mauersberger, D., Rother, K., and Rother, U. (1964). *Ann. N.Y. Acad. Sci.* **120**, 26.

von Krogh, M. (1916). *J. Infect. Diseases* **19**, 452.

Ward, P. A. (1967). *J. Exptl. Med.* **126**, 189.

Ward, P. A., and Becker, E. L. (1967). *J. Exptl. Med.* **125**, 1001.

Ward, P. A., Cochrane, C. G., and Müller-Eberhard, H. J. (1965). *J. Exptl. Med.* **122**, 327.

Ward, P. A., Cochrane, C. G., and Müller-Eberhard, H. J. (1966). *Immunology* **11**, 141.

Wardlaw, A. C. (1960). *Federation Proc.* **19**, 76 (abstr.).

Wardlaw, A. C. (1962). *J. Exptl. Med.* **115**, 1231.

Wardlaw, A. C., and Walker, H. G. (1963). *Immunology* **6**, 291.

Weigle, W. O., and Maurer, P. H. (1957a). *Proc. Soc. Exptl. Biol. Med.* **96**, 371.

Weigle, W. O., and Maurer, P. H. (1957b). *J. Immunol.* **79**, 211.

Weinrach, R. S., Lai, M., and Talmage, D. W. (1958). *J. Infect. Diseases* **102**, 60.

Weintraub, R. M., Churchill, W. H., Crisler, C., Rapp, H. J., and Borsos, T. (1966). *Science* **152**, 783.

Wellensiek, H. J. (1965). *Z. Hyg. Infektionskrankh.* **151**, 222.

Wellensiek, H. J., and Klein, P. G. (1965). *Immunology* **8**, 604.

Wellensiek, H. J., Sauthoff, R., and Klein, P. G. (1963). *Pathol. Microbiol.* **26**, 665.

West, C. D., Northway, J. D., and Davis, N. C. (1964). *J. Clin. Invest.* **43**, 1507.

West, C. D., Davis, N. C., Forristal, J., Herbert, J., and Spitzer, R. (1966). *J. Immunol.* **96**, 650.

Wiedermann, G., Miescher, P. A., and Franklin, E. C. (1963). *Proc. Soc. Exptl. Biol. Med.* **113**, 609.

Wiedermann, G., Ovary, Z., and Miescher, P. A. (1964). *Proc. Soc. Exptl. Biol. Med.* **116**, 448.

Willoughby, W. F., and Mayer, M. M. (1965). *Science* **150**, 907.

Wirtz, G. H., and Becker, E. L. (1961). *Immunology* 4, 473.

Yachnin, S. (1965). *J. Clin. Invest.* 44, 1534.

Yachnin, S. (1966). *Immunochemistry* 3, 505 (abstr.)

Yachnin, S., and Rosenblum, D. (1964). *J. Clin. Invest.* 43, 1175.

Yachnin, S., and Ruthenberg, J. (1964). *Proc. Soc. Exptl. Biol. Med.* 117, 179.

Yachnin, S., and Ruthenberg, J. (1965a). *J. Clin. Invest.* 44, 149.

Yachnin, S., and Ruthenberg, J. (1965b). *J. Clin. Invest.* 44, 518.

Yachnin, S., Rosenblum, D., and Chatman, D. (1964). *J. Immunol.* 93, 540.

Regulatory Effect of Antibody on the Immune Response

JONATHAN W. UHR AND GÖRAN MÖLLER

*Irvington House Institute and Department of Medicine, New York University School of Medicine
New York, New York, and Department of Bacteriology, Karolinska Institutet Medical
School, Stockholm, Sweden*

I. Introduction

During the last decade there has accumulated increasing information on the complexity of antibody formation. In man, five major classes of immunoglobulins, distinguished by their H chains, have been described (Franklin, 1964; Rowe and Fahey, 1965; Ishizaka *et al.*, 1966). These classes can each be subdivided on the basis of two different L chains (λ and κ). The cellular systems responsible for the synthesis of these antibodies are highly complex. In general outline, the lymphoid cells destined to produce immunoglobulins are initially derived from the bone marrow and differentiate into immunocompetent cells under the influence of the thymus or other gut associated lymphoepithelial organs. It has been claimed (Warner *et al.*, 1962; Cooper *et al.*, 1966) that in the chicken one population of such lymphoid cells, under the influence of the thymus, is responsible for the immunological reactions underlying transplantation immunity and delayed hypersensitivity, whereas another population regulated by the bursa of Fabricius is concerned with classic antibody formation. However, the interrelationship of the various lymphoid organs, their connection with the reticuloendothelial system, and the cellular and subcellular basis for the formation of the different classes of immunoglobulins are still largely unknown.

Despite its apparent complexity, the immune response represents a predictable series of events, characterized by the sequential appearance of several classes of γ-globulin antibody molecules and the expression of various cell-mediated immune reactions. The amount of antibody formed to a particular antigen in general appears to be restricted over a wide range of antigen dosage and a variety of immunization regimens and as a rule reaches a predictable maximal level, despite continued immunization. The mechanisms that regulate these events so precisely are still largely unknown.

One mechanism has been described for regulating the concentration of IgG in the circulation (Fahey and Robinson, 1963; Humphrey and Fahey, 1961; Olhagen *et al.*, 1963). The mechanism operates by increasing the catabolic rate of IgG when the serum concentration is abnormally increased as, for example, in multiple myeloma or following hyperimmunization. The mechanism appears to be selective for IgG, e.g., increased serum IgG increases catabolism of IgG but not IgA or IgM. Similar mechanisms have not been described for regulation of serum IgA and IgM. Thus, it could be predicted that the part of the γ-globulin molecule that determines increased catabolism is located in the Fc fragment; this prediction was confirmed by Fahey and Robinson (1963). Since this mechanism lacks immunological specificity, i.e., it cannot distinguish one antibody from another, it can only play a minor role at best in the precise regulation of the immune response.

Dubiski and Fradette (1966) have suggested that allotypes on γ-globulin molecules are involved in a feedback system regulating synthesis of γ-globulin having that allotype. By analogy with the findings of Dray (1962) and Herzenberg and Herzenberg (1966), they demonstrated that injection into neonatal rabbits of antibodies directed against their paternal allotypic specificity resulted in inhibition of immunoglobulin synthesis of the paternal type for 3 to 4 months. After this time, injection of very large amounts of normal rabbit serum into the pretreated animals caused a temporary suppression of immunoglobulin synthesis of the corresponding allotype. Although it was suggested that this suppression of immunoglobulin synthesis was caused by recognition of allotypic specificities, other possibilities cannot be excluded, especially since the levels of immunoglobulins of other allotypic specificities were not determined. Thus, the possibility remains that this represents another example of regulation of the serum concentration of γ-globulin by increased rate of catabolism of γ-globulin subsequent to the rapid increase in its concentration caused by the passive transfer.

Two mechanisms have been described, however, both of which may specifically regulate the immune response. These are (1) alteration of antigenic stimulation. Change in dose, route of administration, use of adjuvants, etc., can modify the amounts of antibodies formed, change the sequential appearance of different classes of antibodies, and influence the duration of antibody formation and of immunological memory (see Uhr and Finkelstein, 1967). Further, immunological unresponsiveness (tolerance) is induced by particular regimens of antigen and results in partial or complete depression of immunocompetence and (2) suppression of the immune response by passive transfer of specific antibodies prior to or shortly after administration of antigen.

Although there are operational similarities between induction of tolerance and antibody-induced suppression of the immune response (suppression), they appear to be differentiated by the status of the immunological system of the animal. Thus, transfer of lymphoid cells from tolerant animals to immunologically incompetent recipients does not transfer specific immunological responsiveness (Medawar, 1960). In contrast, lymphoid cells from recipients treated with suppressive doses of antibodies are competent to initiate an immune response against the corresponding antigen after transfer to incompetent animals (Möller, 1964a). It would appear, therefore, that unlike tolerance induction, suppression does not directly affect the potential antibody-forming cells to the specific antigen.

It has been known for over half a century that mixing antigen with excess antibody can suppress the antibody response. These older studies were aimed at finding methods of immunization against infective agents or toxic substances. Thus, as early as 1892, Van Behring and Wernicke (1892) demonstrated that diphtheria toxin–antitoxin mixtures could be injected into domestic animals in sufficient amounts to stimulate production of immunity without harming the recipients. Similar experiments were performed by Nikanoroff (1897). Theobold Smith (1909) first emphasized the role of excess antibody in decreasing the immune response after antigen injection. In 1909, he wrote ". . . that an active immunity lasting several years can be produced in guinea pigs by the injection of toxin–antitoxin mixtures, which have no recognizable harmful effect either immediate or remote. These studies also show that mixtures which produce local lesions and which, therefore, contain an excess of toxin produce a much higher degree of immunity than the neutral mixtures, and that an excess of antitoxin reduces the possibility of producing an active immunity, and may extinguish it altogether." These early observations were extended during the following several decades to a number of antigens in several species particularly by the English workers, Glenny and Sudmersen (1921), but also by Buxton and Glenny (1921), Ramon and Zoeller (1933), Otten and Hennemann (1939), Regamey and Aegerter (1951), Kalmanson and Bronfenbrenner (1943), Nagano and Takeuti (1951), and Barr et al. (1950).

The possibility that actively formed serum antibody can act as a "feedback" mechanism was suggested by studies indicating that passively administered antibody injected as long as 5 days after immunization was still competent to inhibit the antibody response (Uhr and Baumann, 1961a). The term feedback was used in its operational sense only, without implication as to either the underlying mechanism or to a possible analogy with other feedback systems, such as those that have been described in bacteria. Studies of antibody formation to sheep red blood cells (SRBC) by Rowley and Fitch (1964) and by Möller and Wigzell (1965) have added further significant evidence for a possible regulatory role by serum antibody, in particular as these latter reports were concerned with the effect of passively administered serum antibody on antibody-forming cells utilizing the agar plaque technique (Jerne and Nordin, 1963).

The purpose of this paper is to review recent studies concerning a possible regulatory role of antibody on the immune response. For convenience of presentation, the suppressive effects of antibody on the levels of circulating antibody and the number of antibody forming cells will be discussed separately. Immunological enhancement of tissue grafts, which depends to a major extent upon the regulatory effect of antibody on the development of the immune response is discussed briefly. For a more comprehensive discussion of this topic see Kaliss (1958, 1962), Winn (1959), Snell (1963), Hellström and Möller (1965), and Möller and Möller (1966).

II. Suppression of Antibody Formation by Passively Administered Antibody

A. PRIMARY RESPONSE

1. Specificity

Based on the finding that passively administered horse diphtheria antitoxin completely suppressed antitoxin formation in the guinea pig, whereas a similar amount of nonantitoxic horse γ-globulin did not (Uhr and Baumann, 1961a), it was suggested that suppression was immunologically specific.

A high degree of specificity of antibody-mediated suppression was observed in studies of the immune response to H-2 isoantigens in mice (Möller, 1963a). It was demonstrated that passive transfer of isoantibodies, produced in one strain (A·CA) against a H-2 incompatible strain (A), into A·CA hosts completely suppressed active antibody synthesis to simultaneously transferred strain A cells. In this case the antiserum contained antibodies against *all* major foreign antigenic determinants of the incompatible cells. However, if an antiserum (C3H anti-A), which was directed against only some of the antigenic determinants of the injected strain A cells, was used in the same test system, it was found that complete suppression of active antibody synthesis occurred only with regard to the antigens against which antibodies were present, whereas a normal immune response was observed against "free" antigenic determinants present on the same incompatible cells. Thus, in this system, antibodies suppressed the immune response only to the corresponding antigens, whereas antigenic specificities on the same cells against which antibodies were not present stimulated specific antibody synthesis. The simultaneous occurrence of suppression and immunity after the injection of cells containing multiple antigenic determinants indicates a high degree of specificity.

Wigzell (1966a) studied suppression of the primary 7 S agglutinin response of mice immunized to both sheep and chicken red blood cells. He demonstrated that antibody suppression was specific for the corresponding antigen.

Benacerraf and Gell (1959) investigated the specificity of suppression by antibody to two different antigenic determinants on the same molecule. Guinea pigs were injected with picrylated bovine γ-globulin (BGG) conjugates complexed to rabbit antibody specific to either the hapten or the protein carrier. Suppression of antihapten antibody formation only occurred when antibody to the hapten was used in the immunizing complex. The demonstration that suppression can have immunological specificity implies that interaction with antigen is the first step in suppression. The only mechanism known at present by which an antibody molecule in the serum can recognize antigenic specificity is by combining with specific antigen. No other mechanism is known by which serum antibody molecules can mediate information for the suppression of antibody of a single specificity only, without affecting the other antibody specificities being synthesized. The findings of Gell and Benacerraf (1959) and of Möller (1963a) further suggest that combination of antibody with antigen may not be sufficient to inhibit all antigenic determinants of the molecule (cell), since suppression was restricted to those determinants "covered" by the specific antibody.

The many examples of the specificity of suppression do not exclude the possibility that nonspecific suppression may occur under certain circumstances. Thus, it has been demonstrated (Pokorna and Vojtiskova, 1966) that repeated injections of large amounts of homologous or heterologous normal serum (15–35 ml. per animal) can diminish the frequency of induction of experimental autoimmune aspermatogenesis and can prolong homograft survival. This particular effect on immunization appears different from specific suppression which can be obtained with relatively small amounts of passive antibody.

An apparent lack of specificity of antibody suppression has also been demonstrated with human blood group substances. It is a well-established fact that Rh-negative mothers with Rh-incompatible pregnancies become immunized against the fetus more often if mother and child are ABO compatible than if they are incompatible (Levine, 1943). This has often been referred to as the protective effect of ABO incompatibility on isoimmunization against Rh antigens. Although various explanations have been suggested for this phenomenon, it is now accepted that the presence of natural isoantibodies in the mother against the A or B blood group substances of the infant's red cells is responsible for the effect (Stern et al., 1956). Experiments in humans have demonstrated that passive transfer of antibodies against A or B blood groups to Rh-negative recipients of Rh-positive red cells containing the corresponding A or B group suppressed antibody synthesis against the Rh antigens (Stern et al., 1956). Thus, it would appear that antibodies against one antigen nonspecifically suppressed the immune response against a different antigen present on the same cells. Although these findings may be taken as an argument against the specificity of antibody suppression, they are most likely caused by a mechanism different from that usually involved in suppression. Thus, it seems possible that red cells coated with antibodies against A and B blood groups are localized differently compared to nonsensitized cells, and may be preferentially taken up in organs such as the liver (Mollison, 1951), where they are less likely to stimulate an immune response against the Rh complex.

2. *Ratio of Antibody to Antigen*

Studies using the diphtheria toxoid–antitoxin system indicate that not all antigenic sites on a molecule have to be bound by antibody for suppression to occur (Uhr and Baumann, 1961a). For example, diphtheria toxoid–rabbit antitoxin complexes[1] having an average molecular composition of $Ag-Ab_3$ are as effective in suppressing antitoxin formation in guinea pigs as complexes with average molecular composition $Ag-Ab_5$. Since a single molecule of diphtheria toxoid is known to have a minimum of 6 to 8 sites that can combine simultaneously with antitoxin (Pappenheimer et al., 1940), less than half the available antigenic sites of this molecule have to be bound to antitoxin for effective suppression. Since complexes containing smaller ratios of antitoxin to toxoid were not investigated, it cannot be excluded that as little as one antibody molecule per molecule of toxoid may suffice to suppress antibody formation.

Dixon et al. (1967) studied the ratio of rabbit specific antibody to keyhole limpet hemocyanin (KLH) necessary to suppress antibody formation in rabbits. They observed that the same minimal amount of intravenously administered passive antibody was required to inhibit the immune response to 2 mg. of KLH given intravenously or 2 μg. given in adjuvant. They suggested that the immunogenic stimulus in each situation was similar and that $<1/1000$ of the intravenous dose was engaged in stimulating the immune response. Efficient suppression of antibody synthesis to KLH in adjuvant by anti-KLH, also in the adjuvant, was obtained only with antibodies in extreme excess (average molecular composition $KLH-Ab_{550}$), i.e., an amount capable of saturating all available antigenic sites *in vitro*. Under the conditions studied, it took relatively more antibody to suppress anti-KLH synthesis in rabbits than to inhibit formation of diphtheria antitoxin in guinea pigs.

[1] The commonly used terms, antigen and antibody excess, refer to the status of the supernatants only of antigen–antibody mixtures. For example, in extreme antigen excess in the toxoid–rabbit antitoxin system, the average molecular composition of the complexes is To_2AT; in extreme antibody excess, $ToAT_{6-8}$; in slight antigen excess, $ToAT_2$; and in slight antibody excess $ToAT_3$. There is a continuous spectrum, therefore, of complexes containing increasing ratios of antibody to antigen as the ratio of antibody to antigen in the admixture rises. The transition, therefore, from antigen to antibody excess can reflect a relatively minor change in the average molecular composition of the complexes.

There is an apparent discrepancy between the results with toxoid, suggesting that suppression can be achieved even though several antigenic determinants are left "uncovered" by antibody, and the findings of Benacerraf and Gell (1959) and of Möller (1963a), cited previously, indicating that antibody must be specific to the determinant in question for suppression of antibody to that determinant to occur. As yet these differences cannot be explained. There is little known about the spacing, accessibility, and specificity of the different antigenic determinants on diphtheria toxoid molecules. It is possible, for example, that hapten groups on a heavily conjugated and, thus, slightly denatured protein carrier may be more accessible to potential antibody-forming cells than the antigenic determinants on toxoid molecules. This could explain why a smaller number of antibody molecules is necessary to prevent interaction of toxoid with potential antitoxoid-forming cells. Alternatively, the "processing" in the recipients of the antigens in question complexed to specific antibody may be different.

3. Effect of Immunogenicity of Antigen

In many cases antibody formation to strong immunogens is less easily suppressed than to weak ones; antibody formation is also less easily suppressed to larger doses of immunogens (Uhr and Baumann, 1961a). One exception is the finding that, although diphtheria toxoid is a better immunogen than bovine serum albumin (BSA) in the guinea pig, antibody formation to toxoid is more readily suppressed than is antibody formation to BSA. In man, antibody formation to Rh-incompatible cells, which appear to be weak immunogens, is effectively suppressed by passive antibody after primary immunization has taken place (Clarke et al., 1963, Freda et al., 1964). Under the conditions studied, suppression was more difficult to achieve in rabbits, which are excellent antibody producers, than in guinea pigs, which are relatively poor antibody producers (Uhr and Baumann, 1961a).

4. Effect of Interval between Immunization and Administration of Antibody

The effect on antibody formation of varying the interval between immunization and subsequent administration of passive antibody has been studied in several species with different antigens. In guinea pigs with ϕX phage as antigen, increasing intervals between immunization with antibody injection resulted in progressively decreasing efficiency of inhibition of the primary 19 S antibody response (Finkelstein and Uhr, 1964). Analogous results have been obtained with *Salmonella* flagella in rats (Horibata and Uhr, 1967) and SRBC in rats and mice (Rowley and Fitch, 1964, Möller and Wigzell, 1965). In contrast to these observations, Dixon et al. (1967) found that the efficiency of antibody suppression against KLH in rabbits increased markedly if antibody administration was delayed until 1 day after immunization. At this time, a 90% suppression could be achieved with 5% of the amount of antibody that would react at equivalence in vitro with the amount of antigen that was injected. By ^{131}I labeling of KLH, it was found that most of the antigen was removed from the circulation of normal rabbits during the first 24 hours and less than 1% remained in the whole animal at day 4. These authors suggested, therefore, that the majority of antigen was catabolized during the first day; further, that this process was irrelevant to antibody synthesis and that the minute relevant portion of antigen remained accessible to the passively transferred antibody (although not necessarily in the circulation). Further evidence for this interpretation was provided by the fact that only a few per cent of passively administered ^{131}I-labeled antibody was removed from the circulation by antigen injected 1 day previously, suggesting that only a minute fraction of intravenously injected KLH acts as immunogen (McConahey and Dixon, 1967).

In contrast to the kinetics of antibody synthesis against previously discussed antigens, 19 S and 7 S antibody to KLH in rabbits is not usually detected until 8 to 9 days after primary immunization, and peak titers are not reached until approximately 3 weeks after immunization (Dixon et al., 1966). Meaningful comparisons between the experimental results of these systems is therefore difficult.

B. Preparation for a Secondary Response (Priming)

It is considerably more difficult to inhibit priming than the primary antibody response (Uhr and Baumann, 1961b). For example, doses of antitoxin that partially inhibit primary immunization of rats, guinea pigs, and rabbits usually do not inhibit sensitization for a vigorous secondary antitoxin response. However, using a sufficiently avid antitoxin suppression of priming can occur. It seems likely, therefore, that less antigen is necessary for priming than for detectable antibody formation. Thus, doses of "free" antigen that are insufficient to stimulate detectable antibody formation can, nonetheless, prime animals by whatever mechanism is involved in this process (Salvin and Smith, 1960a).

With certain regimens of immunization, passive antibody can increase the extent of priming. For example, injection of rats with toxoid–antitoxin mixtures appears to sensitize for a greater secondary response than if free toxoid is given alone (Uhr and Baumann, 1961b). Possible mechanisms for the increased immunizing capacity of specific complexes observed under certain conditions is discussed in Section III.

C. SECONDARY RESPONSE

Suppression of this response is also markedly less effective than suppression of primary antibody formation. This is true for diphtheria toxoid in guinea pigs (Uhr and Baumann, 1961b) and SRBC in rats or mice (Rowley and Fitch, 1964; Morris and Möller, 1967). However, suppression can be demonstrated by appropriate experimental techniques (Uhr and Baumann, 1961a). For example, rabbit antitoxin–toxoid complexes are relatively ineffective in eliciting the secondary antitoxin response in "primed" guinea pigs. Also excess horse antitoxin injected several days after secondary immunization of guinea pigs with toxoid in complete adjuvant can partially suppress the secondary antitoxin response.

A complete suppression of detectable serum antibody synthesis in the secondary response can be achieved with H-2 isoantigens in mice: passive transfer of antibodies 15–20 days after primary immunization entirely suppresses the secondary response against simultaneously injected antigenic cells (Möller, 1963c, 1966).

Thus, the mechanism of suppression of the primary and secondary responses is probably similar. The quantitative differences in case of suppression of primary and secondary responses are most likely related to the fact that smaller doses of immunogen are sufficient to stimulate antibody formation in primed than in nonimmunized animals. This is readily understood by considering the progressive increase that occurs in the average binding affinity of serum antibody with time after immunization. Thus, Eisen and Siskind (1964) have shown that the average binding affinity of antibody to dinitrophenyl (DNP) in rabbits can increase 10^4-fold during immunization. This alteration probably reflects concomitant changes in the cell population that produces antibody, i.e., the primed animal would have a population of cells capable of producing antibody of considerably higher binding affinity than those of a previously nonimmunized animal. It is also likely that these antibodies constitute the specific cellular receptors which interact with antigen and give rise to the proliferation and differentiation of cells which underlie the secondary response. The outcome of antibody suppression in the secondary response would thus depend upon the location and relative concentrations and avidities of the host antibody receptors as well as similar properties of the passively transferred antibody. With sufficient amounts of high-avidity antibodies, suppression of the secondary response would be expected.

D. Suppression by Different Classes of Antibody

Smith and Eitzman (1963) found that human infants immunized with *Salmonella* antigens ordinarily formed IgM antibodies. However, if they had received specific 7 S antibodies transplacentally from their mothers, they failed to do so. Sahiar and Schwartz (1964, 1965) studied the antibody response in rabbits immunized to BGG. Treatment with 6-mercaptopurine reduced drastically the IgG response and prolonged the IgM response. Passively administered, specific IgG antibodies terminated IgM anti-BGG synthesis. This finding was interpreted as suggesting that IgG synthesis may terminate IgM synthesis in the normal antibody response. In the ϕX–guinea pig system, it was shown that IgM antibody obtained 5 days after immunization was less effective in suppression than IgG antibody obtained from hyperimmunized animals (Finkelstein and Uhr, 1964). For example, IgM antibody injected at the same time as ϕX suppressed the IgM response at 1 week, but when injected 2–3 days after ϕX immunization, it had no detectable effect on the primary antibody response. In contrast, IgG antibody from hyperimmune animals injected 2–3 days after ϕX immunization was highly effective in suppressing antibody formation; it entirely prevented the primary 7 S response, the development of 7 S immunological memory, and, in addition, partially suppressed the IgM response at 7 days. Möller and Wigzell (1965) studied the suppressive effect of 7 S and 19 S antibody (separated from the same antiserum) against SRBC. Based on the capacity to hemagglutinate, 19 S antibody was far less effective than 7 S in suppression. As they emphasized, however, 19 S antibody is at least several hundred-fold more effective than 7 S antibody on a molecular basis in causing hemagglutination, so that a markedly different number of antibody molecules was probably compared. Because larger amounts of 19 S antibody molecules did cause suppression, the authors suggested that the discrepancy in findings may be of a quantitative nature. Möller (1966) also compared the suppressing ability of 19 S and 7 S antibody directed against H-2 antigens under experimental conditions that accounted for the shorter half-life of the 19 S antibody and observed that only 7 S was efficient. Others have suggested that there is no difference in capacity to suppress between γM and γG antibodies. Rowley and Fitch (1964) have shown that 19 S antibody obtained 1 week after immunization can prevent primary antibody formation to SRBC in rats. Pearlman (1966) studied the suppressive capacity of rabbit antibody to SRBC using 19 S and 7 S antibody of equivalent combining activity and did not find differences in effectiveness of suppression.

There are several difficulties inherent in experiments comparing the effectiveness of suppression of two classes of antibodies. (1) Differences in the capacity of different classes of antibody to perform a particular antibody function, such as agglutination, hemolysis, precipitation, or neutralization of infectivity, makes it difficult to compare the quantities of the two antibody types. (2) Differences in binding affinity of the antibodies may affect suppression. Walker and Siskind (1967) have shown that the ability of an antiserum to suppress is related to the affinity of the antibody for the specific antigenic determinant (ϵ-DNP-L-lysine), high affinity antibody being capable of causing suppression at far lower concentrations than low affinity antibody. Differences in affinity must be taken into account, therefore, when antibody classes such as 19 S, usually obtained early in immunization, and 7 S, usually obtained late in immunization, are compared. Although early 19 S antibody to ϕX has a relatively high avidity when compared to early 7 S antibody (Finkelstein and Uhr, 1966), early 19 S antibody is less avid than hyperimmune 7 S antibody. (3) Differences in specificity of the two classes of antibody may exist. In the one system, in which specificity has been critically studied, 19 S and 7 S have been shown to have similar specificity (Bauer, 1963). However, this does not exclude a more complex situation when other antigens and immunization procedures are used. One meaningful way to compare two classes of antibodies with regard to suppression would be to administer similar numbers of molecules of the two classes of purified antibody having the same immunological specificity obtained from the same serum. Such experiments have not been performed as yet.

The above findings suggest that in the course of immunization significant suppression of antibody formation to certain antigens, such as ϕX and KLH only occurs after a sufficient amount of antibody of high binding affinity has been formed. The question immediately arises whether primary 19 S or 7 S antibody formation is a factor in the termination of 19 S synthesis. The relative importance of antibody suppression is likely to vary with the antigen–host system. Antibody suppression does not appear to be a necessary or even a major mechanism for the termination of 19 S synthesis against ϕX and KLH. This conclusion is derived from the facts that (1) 19 S synthesis against ϕX can terminate at the expected time without 7 S antibody formation (Uhr and Finkelstein, 1963), (2) the concentration of either 19 S or 7 S antibody usually present 1 week after immunization with ϕX is relatively ineffective in terminating synthesis even when injected several days before 19 S synthesis normally terminates (Finkelstein and Uhr, 1964), and (3) large amounts of "primary" antiserum against KLH in rabbits do not effectively inhibit either the primary 19 S or 7 S antibody response to KLH (Dixon *et al.*, 1967).

In contrast to the results with ϕX and KLH, circulating antibody formed during the primary response to other antigens may be a major factor in terminating antibody synthesis. The most striking example is that described by Britton and Möller (1966) regarding antibody formation to a lipopolysaccharide antigen (from *Escherichia coli*) in mice. In this system a single injection of the lipopolysaccharide results in cyclical variations of the serum concentration of the 19 S antibody (no detectable 7 S antibody is formed in the primary response for long periods of time), with differences between the highest and lowest serological titers of about 8 $\log_2$ dilution steps. Most likely, high 19 S serum antibody levels resulting from immunization stop further 19 S synthesis, and 19 S antibody levels then fall rapidly. When the titers have decreased sufficiently, synthesis is again initiated by "free" antigen. The difference between these results with polysaccharides and those reported with ϕX and KLH is probably related to the inability of the host to metabolize efficiently the polysaccharide antigen studied (Britton *et al.*, 1967; Möller *et al.*, 1967). Thus, when catabolic mechanisms cannot be brought into play, serum antibody may be an important mechanism for rendering immunogen unavailable for further stimulation of antibody formation. Also, 7 S antibody may have an important function as a terminator of the primary immune response to SRBC, as revealed by transfer of immune spleen cells (see Section V).

We stress that antibody synthesis may be terminated by a variety of factors other than suppression by antibody, such as depletion of antigen or a self-limited survival time of antibody-producing cells.

E. SUPPRESSION BY ANTIBODY FRAGMENTS

A direct effect of antibody on antibody-forming cells may be effectuated by the Fc fragment of antibody, since this fragment, which does not contain the specific combining sites is essential for many biological functions of antibody, such as complement fixation (Taranta and Franklin, 1961), combination with skin (Franklin and Ovary, 1963), and passage across the placenta (Brambell *et al.*, 1960). Tao and Uhr (1966) studied the capacity of antibody fragments to inhibit antibody formation and found that pepsin-digested antibody was highly effective. Compared to intact antibody with regard to neutralizing capacity, the effectiveness of pepsin-digested antibody was slightly reduced. This may be caused by the altered metabolism of pepsin-digested antibody (Spiegelberg and Weigle, 1965) and/or to minor changes in the configuration of the antigen-combining sites induced by the digestion process. Fitch and Rowley (1967) found that equal suppression of antibody formation to SRBC in rats was produced by rabbit 7 S γ-globulin and by 5 S $F(ab')_2$ fragments produced by pepsin digestion of 7 S γ-globulin. In the KLH system, univalent 3.5 S Fab fragments showed slightly more than one-half the inhibitory effect demonstrated by an equivalent amount of 5 S $F(ab')_2$ fragments (Dixon, 1966). These results further support the concept that the suppression is caused by interaction of antibody with antigen necessary for immunization.

III. Increase of Antibody Formation by Complexing Antigen with Antibody

Complexing antigen with antibody does not always lead to suppression of antibody formation (Downie et al., 1948; Olitzki, 1953; Park, 1922). For example, intravenous injection into mice of BSA does not usually result in its immune elimination or to formation of excess serum antibody. In contrast, a similar injection of BSA complexed previously with a relatively small amount of specific homologous antibody leads to accelerated elimination of the complex and to the formation of excess serum antibody (Terres and Stoner, 1962; Terres and Wolins, 1959, 1961). Similarly, it was found that injection of tetanus or diphtheria toxoid into colostrum-deprived piglets (Locke et al., 1964; Segre and Kaeberle, 1962) did not lead to detectable antitoxin formation. However, injection of a mixture of toxoid with trace amounts of specific horse antitoxin stimulates specific antitoxin formation in such piglets. Möller and Wigzell (1965) have shown that injection of small amounts of 19 S antibody to SRBC along with SRBC may increase the number of 19 S hemolysin-producing cells. Walker and Siskind (1967) observed that very low concentrations of high-affinity antibody, but not low-affinity antibody, increased the expected magnitude of antibody formation to DNP in rabbits. This adjuvant effect of passive antibody may not be specific. Pearlman (1966) observed an increased antibody response in rabbits to *both* determinants of a DNP–SRBC conjugate if small doses of antigen were mixed with excess antibody to SRBC. With large doses of antigen, the mixture with excess antibodies resulted in suppression of hemolysin formation. These results, taken together with those discussed earlier in this review, indicate that depending upon the particular experimental conditions employed, two opposite effects can result from the binding of antigen by antibody: specific suppression and stimulation. Possibly the latter effect is nonspecific and may be caused by the capacity of antibody to change the localization of antigen (see review by Campbell and Garvey, 1963).

The physical state of the immunogenic particle is known to play a profound role in determining the pattern as well as the peak titers of the resulting antibody response. Thus, it is not surprising that binding of relatively small amounts of antibody to antigen may have an adjuvant effect. To carry the argument further, it is possible that natural opsonins (usually present in low concentrations) (Boyden, 1964) may be essential for the induction of antibody formation to certain antigens. The determination of which effect of antibody will predominate probably depends on many factors, but if large amounts of antibody with high binding affinity are employed, suppression will usually occur.

IV.   Effect of Antibody on Metabolism of Antigen

A.   Effect of Antibody on Removal of Antigen

Since the suppressive effect of antibody on the immune response appears to depend on the interaction between antibody and antigen, it is important to consider the effect of antibody on the metabolism and biological stability of antigen. This subject has been extensively covered in recent reviews (Humphrey, 1960; Campbell and Garvey, 1963) and only a few facts will be summarized here. Although the immune status of the recipients is of primary importance in determining the fate of injected antigen, other factors are also significant, such as the physical nature (soluble, particulate) of the antigen, its biochemical properties, and the route of administration.

Soluble protein antigens are removed from the circulation in an exponential fashion during the first days after injection, and the rate of elimination is usually increased when serum antibody at sufficient concentration appears (Dixon et al., 1952). When antibodies are present at the time of injection, the antigen is usually removed rapidly, although exceptions have been found with certain antigens (Möller, 1963d; Erickson et al., 1964).

Part of the antigen removed from the circulation may be catabolized and excreted. For example, it has been demonstrated that labeled protein antigens are catabolized and that (nonprotein-bound) label is excreted in the urine. This occurs faster in immunized than in normal recipients (see review by Campbell and Garvey, 1963).

The antigen removed from the circulation is usually deposited in the liver and the spleen. Clearance of antigens by the liver or the spleen is increased by the presence of specific antibodies. The relative magnitude of antigen localization between these two organs varies with the species injected, the antigen and, also, the presence of specific antibodies.

Experiments with labeled antigens have demonstrated a prolonged retention of antigen in various organs. With protein antigens it has been proposed by several authors on the basis of persistence of radioactivity that the injected protein remains in the spleen and the liver for extended periods. Analogous results were obtained with polysaccharide antigens using different techniques.

In many cases the physical properties of the retained antigen have been studied and it has been possible to establish that retained radioactivity remains attached to protein and may also be associated with the original antigen, e.g., labeled fragments of protein from liver isolated several weeks after injection of ^{35}S-labeled antigen specifically inhibited precipitation reactions between native antigen and antiserum (Garvey and Campbell, 1956). In certain cases it has been demonstrated that the retained antigen is complexed or associated with various cellular constituents, such as lysosomes (Ada and Williams, 1966), ribonucleic acid (RNA) in the case of certain protein antigens, or deoxyribonucleic acid (DNA) with some polysaccharide antigens (see Campbell and Garvey, 1963).

The cellular localization of injected antigen has been studied by different techniques, such as tracing of the antigenic molecules with fluorescent antibodies or by the use of radioactive isotopes and radioautography. These studies have been concentrated as a rule on lymph node and spleen tissue. It has been conclusively demonstrated that protein and polysaccharide antigens are localized in the phagocytic cells of the lymphoid tissues and concentrated to the primary follicles of lymph nodes and spleen. It has been suggested that there is a correlation between the degree of immunogenicity of an antigen and the extent to which it localizes in the follicle (Nossal et al., 1964). Thus, nonantigenic materials were not as a rule found to localize in the follicles, whereas strong antigens, such as bacterial flagella showed marked follicular localization. Weak antigens, such as human serum albumin (HSA), were less intensely localized to the follicles. Using ^{125}I-labeled bacterial flagella, it was shown by Nossal and collaborators (Nossal et al., 1964; Ada et al., 1964, Miller and Nossal, 1965; Nossal et al., 1965b) that macrophages present at the periphery of the lymphoid follicles in immune animals possessed far greater capacity to take up the antigen than corresponding cells in nonsensitized animals. Radioautographs of lymph node sections from animals killed while forming large amounts of antibodies showed that two systems of macrophages were responsible for the trapping and retention of antigen, namely those in the lymphoid follicles and those lining the medullary sinuses. In the medullary sinus, radioactive antigen was present intracellularly in macrophages whereas in the follicles radioactivity was associated with the surface of dendritic processes of reticular cells (Mitchell and Abbot, 1965; Nossal, 1966). It was suggested that this latter localization may be particularly relevant for the immune response since such antigen-containing cells were in close association with lymphoid cells and, thus, made possible their stimulation. Such surface antigen would presumably also be available for antibody binding and might represent the site where suppression occurs.

B. ROLE OF ANTIGEN CATABOLISM IN THE IMMUNE RESPONSE

The influence of retained antigen on the immune response is of central importance. This problem has previously been studied by a variety of techniques, summarized by Humphrey (1960) and by Campbell and Garvey (1963), usually based on the ability of organ extracts of various types from antigen-injected animals to sensitize new recipients for a primary or secondary response. The results have varied depending upon the antigen and species under investigation. Certain protein antigens, such as diphtheria toxin, may be antigenic in the lymphoid tissues for 3 weeks, whereas others, such as KLH, are immunogenic only for 3 days. In the latter case, however, organ extracts from animals immunized up to 40 days previously could inhibit specific antigen–antibody precipitation. Analogous results have been obtained with BSA and BGG. Polysaccharide antigens, on the other hand, are known to persist in an immunogenic form for very long periods of time.

More complex antigens, such as SRBC, have also been tested in systems of the above type. It has been found that their immunogenicity, as revealed by transfer tests, decayed rapidly; most of the activity disappeared after 1 day and was virtually gone by 3 days (Franzl, 1962; Franzl and Marello, 1966).

However, the interpretation of tracer experiments and transfer tests of the above type with regard to persistence of antigenicity is uncertain. With tracer techniques the possibility exists that the isotope is dissociated from the antigen or, if this can be excluded, that the antigen detected does not play a role in the immune response. Negative results in transfer tests do not necessarily imply that immunogenicity in the primary recipients has been lost, since the antigen may have been catabolized to such an extent that it cannot function as a complete antigen in a secondary host, even though it may have had the capacity to stimulate an immune response in its original host. Evidence for such a change in antigen is that organ extracts from antigen-injected animals which contained material capable of reacting with specific antibody *in vitro,* failed to sensitize secondary hosts (Garvey and Campbell, 1956).

Another approach, which overcomes these difficulties to some extent, has been used to test the persistence of immunogenicity. This test is based on the ability of retained antigen in the primary host to stimulate an immune response in subsequently injected lymphoid cells. In practice, these experiments were carried out by lethally irradiating recipients of the antigen, repopulating them with syngeneic bone marrow and spleen cells, and then following antibody synthesis against the originally injected antigen. Control experiments showed that the dose of irradiation was sufficient to abolish antibody synthesis by the previously committed lymphoid cells of host origin, and any antibody production which occurs in such animals is assumed to be carried out by the transferred lymphoid cells. Experiments of this general design have been performed with SRBC, bacterial lipopolysaccharide antigens (Britton *et al.,* 1967), and HSA (Mitchison, 1965; Britton and Celada, 1966). Bacterial lipopolysaccharide antigens were found to persist in an immunogenic form for very long periods (70 days) and stimulated a primary **19 S antibody** response in the transferred nonimmune lymphoid cells. SRBC remained immunogenic for 14 days, but lost this capacity after 3 weeks. The possibility cannot be excluded, however, that the failure to obtain a primary response in animals injected with SRBC 3 weeks previously was caused by suppression of the primary response by the previously synthesized 7 S antibodies rather than by decay of immunogenicity. A slightly modified test system was used to study the immunogenicity of HSA: presensitized lymphoid cells were transferred to nonlethally irradiated hosts, which had been injected with HSA at various intervals previously. Since in the doses used, this antigen does not stimulate a detectable primary response, the antibodies appearing after cellular transfer are produced as a consequence of antigen stimulation of a secondary response. By this test the half-life of immunogenicity of HSA *in vivo* was found to be identical with the half-life of HSA in the circulation of mice, namely, 16–17 hours.

The results with lipopolysaccharides are in agreement with previous studies, whereas a discrepancy exists with regard to SRBC between the above findings and those previously obtained with transfer systems using organ extracts (Rowley and Fitch, 1967; Franzl, 1962; Franzl and Marello, 1966). This discrepancy may be caused by the additional catabolism involved in transfer studies as previously discussed.

C. Release of Antigen from Specific Complexes

With certain diphtheria toxoid–antitoxin complexes, antitoxin formation could be suppressed for many weeks, but an immune response eventually developed at a rapid rate (Uhr and Baumann, 1961a). These observations suggest the possibility that formation of serum antibody at high rates may be caused by stimulation of the immune mechanism by "free" antigen which has dissociated from specific complexes. This possibility has been tested (Uhr and Baumann, 1961a) by using diphtheria toxin, a biologically active protein, which can be detected in minute amounts if released from specific complexes. As a first step it was shown that toxigenicity was not lost when toxin was released from specific precipitates *in vitro* by treatment of the precipitate with a large excess of toxoid. It was demonstrated that well-washed toxin–antitoxin complexes, which had been prepared in the zone of antitoxin excess and which did not produce inflammation when injected into the skin of rabbits, nevertheless could cause paralysis and death of guinea pigs and rabbits 2–4 weeks after intravenous injection. It would appear, therefore, that dissociation of antigen can occur from specific complexes. Moreover, the pattern of dissociation paralleled that suggested by immunization experiments. For example, toxoid–guinea pig antitoxin complexes did not suppress as well as toxoid–rabbit antitoxin complexes in guinea pigs; both suppressed equally well in rabbits. Similarly, toxin–guinea pig antitoxin complexes, but not toxin–rabbit antitoxin complexes caused death in guinea pigs, whereas both types of complexes caused death in rabbits.

The kinetics of release *in vivo* of toxin from toxin–antitoxin complexes was studied by injecting a large excess of horse antitoxin at various times after injection of the complexes and observing when protection occurred. It was found that horse antitoxin, injected 1 hour after toxin–guinea pig antitoxin precipitates, completely prevented the expected late intoxication. Horse antitoxin injected even 5 hours later provided substantial, but not complete protection. Since the visible precipitates used in these experiments were presumed to be cleared from the circulation within minutes, the release of toxin from the specific complexes must have occurred after their uptake by the reticuloendothelial system.

The means by which antigen escapes from the specific complexes is not known. Antigen may dissociate from intact complexes or the antibody molecules may be digested, thereby exposing antigenic determinants. In any event, the amount of antigen that dissociates from an antigen–antibody complex of the type described above appears to be relatively small;

in the case of toxin–antitoxin complexes perhaps as little as 0.001 μg. of toxin is released in guinea pigs receiving 3 μg. of toxin–guinea pig antitoxin (Uhr and Baumann, 1961a).

V. Suppression of the Number of Antibody-Synthesizing Cells by Antibody

Although the principles for antibody-induced suppression of the immune response have been elucidated to a large extent by studies of serum antibody, the development of techniques for the detection of individual antibody-producing cells *in vitro* have made possible a more detailed study of the kinetics of suppression and of the mechanisms involved.

A. CHARACTERISTICS OF THE CELLULAR IMMUNE RESPONSE

In the direct agar plaque technique, as described by Jerne and Nordin (1963), Ingraham and Bussard (1964), and Sterzl and Mandel (1964), the number of cells producing antibody to SRBC [plaque-forming cells (PFC)] starts to increase after a latent period of approximately 24 hours and thereafter increases exponentially, reaching a peak 4–5 days after immunization. Subsequently, there is a rapid decline of the number of PFC. The number of PFC is paralleled by 19 S antibody titers (Jerne and Nordin, 1963; Jerne *et al.*, 1963; Wigzell *et al.*, 1966). Although extrasplenic antibody synthesis occurs in lymph nodes and peripheral blood, it is quantitatively of minor importance and shows the same kinetics as does splenic production. Serum 7 S antibody is detected by 5 to 7 days after immunization as a rule and increases during the first 12–20 days. Nonimmune animals always show a background of PFC varying as a rule between 20 and 200 PFC per spleen with marked quantitative variations between different strains. These cells are completely absent in germfree piglets raised on a synthetic diet (Sterzl *et al.*, 1966). Nevertheless, such animals respond normally to antigen injection. Thus, the background PFC may not be precursors of the immune PFC (Sterzl *et al.*, 1966). Furthermore, the background levels of PFC cannot be suppressed by passive antibody (Wigzell, 1966c), although immune PFC are easily inhibited as described later. The effect of various antigen doses on the development of PFC has been studied by several authors. An optimal dose results in an exponential increase of the number of direct PFC with a doubling time of 5 to 7 hours, and the peak is reached 4–5 days after immunization. Decreasing doses of SRBC result in a lower number of direct PFC at the peak and a gradual shift of the peak to later times. Thus, the rate of increase of the number of PFC decreases with smaller antigen doses. Qualifications of this scheme will be made below in connection with studies on cellular 7 S synthesis.

The role of cell division in the development of PFC is not settled as yet. It has been demonstrated by Jerne (1966) that about 25% of the PFC from immune animals incorporate tritiated thymidine after incubation with the isotope for 1 hour *in vitro*. However, Sterzl *et al.* (1966) investigated the incorporation of tritiated thymidine into PFC by labeling *in vivo* and found a substantial uptake only if the thymidine was given between 3 and 4 days after immunization. Prior to this only a small fraction of the PFC was labeled. He suggested that the developing PFC were not multiplying until 3 to 4 days after antigen injection.

It should be stressed that the increase in cellularity of the lymphoid organs after antigen injection cannot be accounted for to a major extent by specific antibody-producing cells, since the proportion of such cells is low (less than 1 in 1000 at the peak of the response).

It seems clearly established that the PFC do not belong to a single clone of antibody-producing cells but are derived from multiple clones, which are recruited at different times after immunization. This conclusion is derived from the demonstrations that the development of PFC occurs in colonies of different sizes in anatomically different sites in the spleen (Playfair *et al.*, 1965; Kennedy *et al.*, 1966).

The agar plaque technique can be used also with antigens other than SRBC. Thus, erythrocytes from other species, such as chickens, cattle, pigs, and rabbits have been successfully employed (Jerne, 1965; Wigzell, 1966a; Playfair *et al.*, 1965; Nakano and Braun, 1966; McBride and Schierman, 1966). Furthermore, various soluble antigens, such as lipopolysaccharides (Landy *et al.*, 1965; Möller, 1965) and haptens (Merchant and Hraba, 1966) have been coated onto SRBC and used for the detection of cellular antibody synthesis.

It has been clearly demonstrated that animals simultaneously injected with two antigens have antibody-producing cells against each antigen in discrete anatomical sites in the spleen (Playfair *et al.*, 1965). This observation was made by injecting nonimmune lymphoid cells into lethally irradiated recipients treated with two different red cell antigens and subsequently localizing the sites of hemolysin-producing cells in the spleen against both antigens simultaneously. Analogous conclusions were reached by a slightly different test system (Celada and Wigzell, 1966): presensitized animals were sublethally irradiated and thereafter received a secondary injection of the two antigens. The spleen colonies developing after irradiation in the recipients were isolated and the presence of PFC against the two antigens determined. Finally, it has been shown that PFC are nonrandomly distributed and that when two different antigens are injected, the nonrandom distribution of PFC to one of them differs significantly from that of the PFC to the other antigen, suggesting clonal distribution of PFC (Nakano and Braun, 1966).

By use of technical modifications, the plaque assay has been adopted to detect cells that produce nonhemolytic antibodies (Dresser and Wortis, 1965; Sterzl and Riha, 1965; Weiler et al., 1965). In the modified method, a "developing" antiserum produced in a foreign or the same species against immunoglobulin of the spleen cell donor is used to cause the appearance of PFC which are not revealed by the addition of complement alone (indirect PFC). By the use of this technique the usual decrease of the PFC after 4 to 5 days does not occur, and the number of PFC is sustained at high levels for a longer period. It is assumed that the PFC detected in this way represent 7 S-producing cells, but this possibility has not been critically demonstrated in most studies. Cross-reactivity between IgM and IgG because of their L chains is expected to occur with the developing sera generally used, and, consequently, the possibility exists that the indirect plaque technique may also detect cells producing IgM of low hemolytic efficiency. Absorption experiments have made this possibility unlikely (Wigzell, 1967) and with highly specific sera to the different immunoglobulin classes it has been possible to detect cells producing IgA, IgG, and other immunoglobulin classes (Papermaster, 1967; Dresser, 1967).

Various attempts have been made to study whether IgM and IgG cells develop from the same cell line or arise in two independent lines. This has been approached by applying the previously described technique to study PFC in spleen colonies appearing in sublethally irradiated recipients being immunized against two antigens (Celada and Wigzell, 1966). It was shown statistically that the distribution of direct and indirect PFC in the different colonies was random. In this test system, sensitization was carried out twice—before and after X-irradiation. It cannot be excluded, therefore, that primary and secondary immune responses had occurred simultaneously and, consequently, that the test revealed the cellular independence of these two responses rather than of the IgM and IgG responses. However, analogous results have been obtained (Möller, 1966) in a primary immune response using the technique described by Playfair et al. (1965) and Kennedy et al. (1966). Each spleen was cut into 50 to 100 pieces, and each piece was studied for direct and indirect clones. It was found that indirect 7 S clones appeared in sites where no direct (IgM) clones were found. However, neither this study nor that of Celada and Wigzell (1966) clarifies the relationship between IgM and IgG synthesis, since the described results may occur whether IgM and IgG are derived from independent cell lines or not. Even if one cell line is responsible for synthesizing both IgM and IgG, it would be expected that colonies would appear containing only one class of antibody; i.e., the cells producing IgM at the moment of detection may synthesize IgG at a later stage and the IgG-producing cells may already have switched. If such a switch occurs rapidly it is unlikely that it would be detected by the above methods.

B. Suppression of the Primary Immune Response

1. Antibody Suppression of the 19 S Response

Demonstration at the cellular level of suppression of the primary immune response has been achieved with passively transferred antibody directed against SRBC and bacterial lipopolysaccharide antigens (Rowley and Fitch, 1964, 1965a,b, 1967; Möller and Wigzell, 1965; Britton and Möller, 1966). In analogy with the previous results at the serological level it was found that large amounts of passively transferred antibody given prior to or at the same time as the antigen completely suppressed the development of 19 S-producing PFC, as well as the morphological changes and the increase of spleen weight that accompany the primary response (Sahiar and Schwartz, 1966; Rowley and Fitch, 1964). Hyperimmune sera were efficient whether absorbed onto the antigen or injected by a separate route (Rowley and Fitch, 1964; Möller and Wigzell, 1965). Suppression has been shown to be immunologically specific (Rowley and Fitch, 1964; Wigzell, 1966a).

Rowley and Fitch (1964) found 19 S and 7 S antibodies to be equally efficient with regard to suppression of cellular antibody production against red cells in rats. In this case, inhibition was compared with sera taken at different times after immunization and not with purified fractions from the same serum. In other experiments (Möller and Wigzell, 1965) purified (Sephadex G-200) fractions from the same serum were used, and it was found that only 7 S antibodies efficiently suppressed the immune response, whereas 19 S was comparatively inefficient. However, 19 S antibodies could also suppress the development of PFC if the ratio of antibody to antigen was increased 100–200 times based on hemagglutination titers (Möller and Wigzell, 1965). Presumably the discrepancies in results are related to the difficulties inherent in comparisons of this type as discussed previously.

In certain experimental systems, antibody-induced suppression of the immune response appears to require very small quantities of antibodies.

For example, suppression of cellular 19 S synthesis against lipopoly-saccharide antigens of *Escherichia coli* can be achieved with the 7 S fraction of hyperimmune sera by using amounts that do not contain detectable antibody as revealed by hemolysis and hemagglutination using red cells sensitized with the antigen (Britton and Möller, 1967).

In agreement with previously reported results on immune responses at the serum level it was found that passively transferred serum anti-bodies could suppress the number of antibody-forming cells even if injected after antigen (Möller and Wigzell, 1965). However, it was found that the degree of suppression decreased with increasing time intervals between the injection of antigen and antibody. Maximal suppression was obtained with antibody given before or at the same time as the antigen, and pronounced suppression could also be achieved with anti-body injected 24 and 48 hours after the antigen injection. The differences between these findings and those obtained with KLH as described previously (Dixon *et al.*, 1967) may be caused by differences in metab-olism of the two antigens or in the kinetics of the subsequent immune responses.

With the plaque technique, it was found that the suppressive effect of antibodies did not make itself manifest during the first 40–48 hours after injection, irrespective of the time interval between antibody and antigen treatment (Möller and Wigzell, 1965). After this latency period suppression was revealed as a termination of the exponential *increase* of the number of PFC. Thus, the actual number of PFC observed at 48 and 72 hours after antibody injection was approximately equal. Conse-quently, suppression was revealed by the termination of the exponential increase in the number of PFC 48 hours after antibody injection. During the subsequent 24 hours the number of PFC remained constant in anti-body-treated animals. Thereafter the number of PFC decreased rapidly and eventually reached background levels. The rate of decrease after antibody treatment was the same as that observed in nonantibody-treated animals after the peak number of PFC was reached 4–5 days after the primary antigen injection. Antibodies given after the peak also decreased the number of PFC in a similar way (Wigzell, 1967). It would appear, therefore, that the sequence of events responsible for the exponential increase or maintenance of the number of PFC, whether caused by cellular multiplication, nonmitotic differentiation, or both, pro-ceeds without interruption for 40 to 48 hours after antibody treatment. After this latent period, the suppressive effect was readily detected. This suppression may be caused by inhibition of the development of new PFC or by a decreased rate of synthesis of antibody in each cell to undetectable levels or by some combination of these effects. It has been suggested, however, that antibodies do not effect the rate of synthesis in committed cells based on the finding that the average size of the plaques was the same in antibody-treated and untreated hosts (Wigzell, 1967).

Horibata and Uhr (1967) observed the effect of injecting passive antibody after antigen on the antibody content of individual antibody-forming cells appearing during immunization. They used bacterial adherence to identify the antibody-forming cells and the micromanipulation techniques of Mäkelä and Nossal (1961) to study the intracellular antibody content. The results showed that popliteal lymph node cells from flagella-immunized rats did not contain detectable antibody 4 days after immunization but did have considerable antibody 6–7 days after immunization. It appears, therefore, that antibody content of cells at this early stage in immunization can reflect the differentiation process of antibody-forming cells. When passive antibody was injected 1 day after immunization, the number of antibody-forming cells detected during the primary response was markedly reduced and their intracellular antibody content was below the level of detection at 6 to 7 days after immunization. The results suggest that interruption of continued antigenic stimulation by passive antibody has markedly slowed entrance of potential antibody-forming cells into the antibody-forming cell compartment ("recruitment") or the rate of differentiation within that compartment.

The possibility exists that antibodies directly suppress the antibody-synthesizing cells. Different opinions have been expressed with regard to this possibility. Thus, Rowley and Fitch (1964, 1967) have suggested that antibodies can interact directly with immunologically competent cells and suppress their ability to produce antibodies subsequent to antigenic stimulation. This was based on the demonstration (Rowley and Fitch, 1964) that normal rat lymphoid cells, which had been brought into contact with humoral antibodies against SRBC in vitro or in vivo, were incompetent to produce antibodies after transfer to irradiated recipients subsequently injected with SRBC. Since the antibody-mediated suppression of the immune response usually shows a stringent requirement for immunological specificity and, thus, appears to be effected by an interaction between antibody and antigenic determinants, they postulated (Rowley and Fitch, 1967) that the potential antibody-forming cells carry on their surface or within themselves certain sites, which are similar to or identical with the antigenic determinants. However, conflicting results obtained by Möller (1964a) indicated that treatment of normal mouse spleen and lymph node cells with humoral antibodies to *Salmonella* H antigens did not suppress the ability of the lymphoid cells to carry out a primary immune response against the corresponding antigen after transfer to irradiated recipients. Analogous findings have been obtained with thoracic duct lymphocytes in rats (McCullagh and Gowans, 1966) and with spleen cells in mice (Wigzell, 1967) using SRBC antigens. The possibility exists that the findings of Rowley and Fitch may be explained by adsorption of antibodies to the nonimmune lymphoid cell population, which most likely contains a proportion of macrophages known to fix antibodies to their surface; these antibodies may have been transferred and exerted a suppressive effect on the induction of the primary immune response by interaction with the injected antigen. This possibility is strengthened by the demonstration that only a small amount of antibody is necessary for suppression of the immune response against certain antigens.

2. *Antibody Suppression of the 7 S Response*

The suppressing ability of passively transferred antibodies on the primary immune response has been studied also with regard to 7 S synthesis (Möller and Wigzell, 1965; Wigzell, 1966a). The PFC which are detected by the indirect method using a heterologous anti-γ-globulin serum will be referred to as 7 S or IgG PFC below (see previous discussion). As discussed previously, it has been repeatedly demonstrated that humoral antibodies injected simultaneously with the antigen results in marked suppression of 19 S and of 7 S production as well. Antibodies injected after the antigen often suppress 19 S synthesis, although the latent period of suppression allows the 19 S titers and the number of PFC to increase for some time. As a rule 7 S production is also completely suppressed by antibodies given during the first 3 days after the antigen. However, with increasing intervals between the antigen and antibody injection, 7 S synthesis becomes more difficult to suppress, and serological studies suggest that detectable suppression of 7 S production cannot be achieved by antibodies given during the early exponential phase (6–7 days after antigen injection) of 7 S synthesis against SRBC (Möller and Wigzell, 1965). However, the serological techniques used for these studies would not detect less than a 50% suppression of the antibody synthesis. It has been generally assumed, however, that 7 S synthesis is much less dependent upon the continuous presence of antigen or requires lower concentrations of antigen for its maintenance than 19 S synthesis (Uhr, 1964; Svehag and Mandel, 1964). Nevertheless the repeated demonstrations of efficient antibody-mediated suppression of 7 S production during the first 3 days after antigen injection clearly indicate that the events leading to 7 S antibody synthesis are antigen dependent at this early stage.

Recent experiments on antibody suppression of 7 S synthesis have been performed with the indirect agar plaque technique (Wigzell, 1966a). It was demonstrated that passive transfer of antibodies against SRBC and chicken red cells from hyperimmune animals caused a marked decrease in the number of 7 S PFC against the specific antigens even when transfer was carried out several weeks after antigen injection. Suppression was observed after a latent period of 48 to 72 hours and required large amounts of passively transfused antibodies for a detectable effect. The suppressing efficiency of the antibodies increased with the number of antigen injections of the donor, and hyperimmune sera were essential for a demonstrable effect. Suppression appeared to be immunologically specific, since antibodies against chicken red cells were ineffective against SRBC and vice versa. Studies on the serum levels of 2-mercaptoethanol-resistant 7 S agglutinins also demonstrated specific suppression of antibody synthesis.

The interpretation of the immunological events leading to suppression of 7 S synthesis remains unclear. It is not possible to differentiate between a suppressing effect of the injected antibodies on antigenic stimulation of 7 S-producing cells or on precursors of these cells. For example, a continuous recruitment of 19 S synthesizing cells may occur during the immune response and these cells may eventually switch to 7 S production; termination of this recruitment would affect 7 S production. Another possibility is that 7 S cells may continually arise from 7 S-"memory" cells which are stimulated by antigen persisting from the primary injection.

Regardless of mechanism, the experimental demonstration of suppression of 7 S synthesis a long time after its initiation suggests that 7 S production is dependent on continuous antigenic stimulation.

3. Feedback Regulation of the Primary Response

It has been suggested (Sahiar and Schwartz, 1964; Möller and Wigzell, 1965) that the termination of 19 S synthesis against SRBC occurring 4–5 days after the injection of antigen may be caused by a feedback suppression of 19 S synthesis by 7 S antibodies, which start to appear at this time. However, other events may also be responsible for abrogation of 19 S synthesis. Thus, the dependence of 19 S synthesis on antigen stimulation suggests that metabolic degradation and/or excretion of the immunogen may terminate 19 S synthesis. This possibility is not supported, however, by the previously described experiments showing that SRBC persist for prolonged periods *in vivo* in an immunogenic form.

Several attempts have been made to study the significance of antibody-induced suppression in the regulation of the primary response. It has been experimentally demonstrated that antibodies synthesized within an animal may suppress the induction of a primary response in the same animal, thus excluding the possibility that suppression is mediated only by passively transferred antibodies (Rowley and Fitch, 1964; Morris and Möller, 1967). Thus, Rowley and Fitch showed that a small sensitizing dose suppressed the response to a subsequent large antigen inoculum. Morris and Möller (1967) inoculated hyperimmune spleen cells into untreated syngeneic animals and at different intervals thereafter injected SRBC into the recipients and tested them for production of PFC and serum antibody titers. It was found that 19 S-producing PFC were markedly suppressed, whereas there was a pronounced stimulation of 7 S PFC. It follows that 7 S antibodies produced by the transferred cells efficiently suppressed initiation of 19 S synthesis in previously uncommitted cells. The transferred 7 S cells themselves reacted with a secondary response, revealed as an increased number of 7 S PFC and 7 S serum antibodies.

To test whether 7 S antibodies produced during the primary response may suppress 19 S synthesis the same test system was used, but the immune spleen cells, which were transferred to syngeneic recipients, were taken from animals immunized against SRBC 5–10 days previously. The antibodies produced by such cells also suppressed initiation of 19 S synthesis.

Strong support for a regulating role of antibody during the primary response has been obtained in another system (Britton and Möller, 1966, 1967). A lipopolysaccharide antigen from *Escherichia coli* was found to stimulate mice to synthesize 19 S only for several months; 7 S antibodies were subsequently detected after hyperimmunization. By means of the agar plaque technique, using sheep red cells coated with the antigen (Möller, 1965), it was found that injection of the antigen in the form of a bacterial vaccine stimulated the production of 19 S PFC which initially followed the same pattern as observed with SRBC. Thereafter, new peaks of PFC appeared at regular 10–15 day intervals, each peak being slightly lower than the preceding one as a rule.

The regular cyclical fluctuations of cellular 19 S synthesis after one antigen injection was ascribed to a feedback suppression of 19 S synthesis by its own product (Britton and Möller, 1967, Möller *et al.*, 1967):

the antigen would stimulate 19 S synthesis in lymphoid cells and the number of antibody-synthesizing cells would increase exponentially during the first 4–5 days. Eventually the titers of 19 S antibodies would reach a sufficiently high concentration to be able to interact with the antigen in such a way as to suppress its ability to stimulate further 19 S synthesis. Since no shift to 7 S production occurred in this system and since both the 19 S-producing cells and the serum antibodies have a short half-life, suppression would be gradually lost. Provided the antigen was still immunogenic it could then initiate a new cycle of 19 S PFC. According to this hypothesis it would be possible to suppress any peak of PFC by passive transfusion with specific antibodies shortly before the expected appearance of the peak. Several experiments of this type have been performed (Britton and Möller, 1967), and it was shown that passively transferred 19 S antibodies were competent to suppress the appearance of PFC for about 10 days, but thereafter PFC started to appear. It seemed likely that the progressive metabolism of antibody diminished its concentration below that sufficient to suppress antigenic stimulation of the immune response.

An essential requirement for this hypothesis is that antigen should persist in an immunogenic form for a long period of time. As discussed above, this has been tested by injecting the lipopolysaccharide antigen into different groups of animals which thereafter were irradiated with 900 r at various intervals and then repopulated with syngeneic bone marrow and lymphoid cells from nonimmune animals (Britton et al., 1967). It was found that the repopulated animals produced antibodies to the same extent as nonirradiated animals, even if the interval between antigen treatment and repopulation was as long as 70 days.

C. SECONDARY RESPONSE

Studies of serum antibody formation have shown that the secondary response may be suppressed by large quantities of passively transferred antibodies, although less easily than the primary response (see Section II). Analogous studies at the cellular level have been performed to a limited extent only. As a rule, the secondary cellular 19 S response is lower than the primary after optimal antigen doses, presumably because the presence of 7 S antibodies interferes with induction of 19 S synthesis. This is strengthened by the demonstration that transfer of hyperimmune spleen cells, producing few 19 S PFC, into irradiated animals injected with antigen (SRBC) at the same time results in a marked increase of 19 S PFC (Wigzell, 1967). Thus the 19 S cells are competent to respond to antigen in an antibody-free host. The number of 7 S PFC increases markedly after challenge even in the presence of antibodies. When immune spleen cells are transferred to untreated syngeneic recipients, which are challenged with SRBC within 1 to 2 weeks, there is a marked increase of indirect (7 S) PFC and a depression of 19 S PFC when measured 4 days after challenge. Since very few 7 S PFC appear in a primary immune response at day 4, it follows that these cells are derived from the transferred immune cells. The suppression of the 19 S PFC suggests that the 7 S antibodies produced by the transferred cells suppressed the development of 19 S PFC.

The outcome of antibody suppression in the secondary response probably depends on the quantity and avidity of antibody in the serum as compared to that associated with the lymphoid cells. Large amounts of highly avid antiserum may compete successfully for antigen with immune lymphoid cells.

VI. Suppression of Delayed Hypersensitivity and Experimental Autoimmunity

Delayed hypersensitivity to protein antigens appears to be less easily suppressed than is either the primary antibody response or "priming" for a secondary antibody response. For example, complexes of toxoid and highly avid rabbit antitoxin injected in complete Freund's adjuvant into guinea pigs can induce delayed hypersensitivity indistinguishable from that induced by "free" toxoid in adjuvant (Uhr et al., 1957). This is revealed by delayed skin reactivity, detectable by day 4, which reaches a maximum by day 7 or 8. Indeed, all attempts to demonstrate suppression of delayed hypersensitivity in this system have failed, despite the use of large amounts of heterologous antitoxins (Uhr and Baumann, 1961a). However, antibody-mediated suppression of induction and expression of transplantation immunity (presumed to be at least partly a manifestation of cell-mediated immunity) has been clearly demonstrated as is discussed in the later section on enhancement.

A possible explanation of the inefficiency of serum antibody in suppressing delayed skin reactivity can be derived from the hypothesis that delayed reactivity is caused by antibody molecules of relatively high binding affinity (Karush and Eisen, 1962; Levine, 1965). If so, delayed hypersensitive cells equipped with such antibody would be able to capture specific antigen, even if the latter were injected in the form of specific precipitates formed with excess, but less avid antibody. No experimental support for this hypothesis has emerged, however. Another possibility is that induction of delayed skin reactivity can be achieved with smaller quantities of antigen than are required to stimulate detectable antibody formation or priming for a secondary response. Minute amounts of antigen may dissociate from antigen–antibody complexes or there may exist sterically available determinants despite binding to antibody, which might suffice to induce delayed hypersensitivity but no other immunological manifestations. The same argument may apply to the capacity of antigen–antibody complexes formed in antibody excess to elicit as well as to desensitize the delayed hypersensitive state (Uhr and Pappenheimer, 1958). Finally, the possibility exists that serum antibodies used for suppression of delayed reactions have a different specificity from those involved in expression of delayed hypersensitivity (Benacerraf and Gell, 1959, 1962; Salvin and Smith, 1960b). If the antibody used in complexes with antigen is only directed against part of the antigenic determinant(s) responsible for delayed hypersensitivity, efficient suppression might not occur.

In analogy with delayed hypersensitivity, many autoimmune conditions have been shown to be mediated primarily by immune lymphoid cells. The effect of serum antibody on the manifestation of experimental autoimmunity has been most extensively studied with experimental allergic encephalomyelitis (AE). This disease can be transferred to otherwise untreated recipients by sensitized lymphoid cells (Paterson, 1960). In addition immune cells can cause damage to myelinated target cells in culture (Berg and Källén, 1962; Koprowski and Fernandes, 1962). Serum antibodies are also efficient against target cells *in vitro* (for reference see Paterson, 1966) although they do not transfer the disease *in vivo*. In the course of the disease there is often an inverse relation between serum antibody and occurrence of AE. Transfer of antibrain serum suppressed AE if given prior to sensitization of the recipients (Paterson and Harwin, 1963).

VII. Immunological Enhancement of Tissue Grafts

Immunological enhancement is defined as the prolonged survival of normal or neoplastic tissue grafts in histoincompatible recipients, which have been pretreated with antibodies directed against the graft or, alternatively, which have been presensitized with tissue of the graft genotype. This phenomenon is paradoxical in so far as the pretreatment of the recipient involves active or passive immunization which would normally be expected to accelerate the rejection of the transplanted graft. Indeed, most tissues grafted into sensitized foreign hosts are rejected in an accelerated fashion. The principle responsible for the induction of enhancement is humoral antibody (Kaliss, 1958). The outcome of an individual experiment is to a large extent determined by the sensitivity of the transplanted cells to cytotoxic antibodies (Gorer and Kaliss, 1959; Möller, 1963b): cells that are resistant to isoantibodies as a rule grow progressively in antibody-pretreated hosts, i.e., exhibit enhancement, whereas sensitive cells undergo accelerated rejection.

Cellular sensitivity to isoantibodies is, to a large extent, determined by the cell's concentration of surface histocompatibility antigens (Möller and Möller, 1962; Winn, 1962), but other properties of the antibodies also influence the results. Thus, it has been shown that 19 S antibodies are more active in cytotoxic tests than are 7 S antibodies in the presence of guinea pig complement, whereas the converse is true in the presence of rabbit complement (Winn, 1965). In agglutination tests, 7 S antibodies are more potent than 19 S (Möller, 1966; Winn, 1965; Andersson et al., 1967). Immunological enhancement can be achieved with 7 S antibodies, whereas 19 S antibodies appear inefficient, even under experimental conditions that take their short half-life into account (Möller, 1966).

The phenomenon of immunological enhancement is to a large extent caused by the ability of antibodies to suppress the immune response (Snell et al., 1960; Brent and Medawar, 1961; Möller, 1963a,c). In addition to the suppressing effect of antibodies on the induction of the homograft reaction, other mechanisms are operative in the enhancement phenomenon. Of particular importance is the ability of antibodies to inhibit the effector mechanisms of the homograft reaction by antagonizing the cytotoxic effect of cell-mediated immune reactions (Möller, 1963c, 1964a; E. Möller, 1965b). A short summary of the phenomenon of immunological enhancement follows, with main emphasis on those aspects of the antibody action that have not been discussed in the previous sections concerned with nonreplicating antigens. For further details see reviews by Kaliss (1958, 1962), Hellström and Möller (1965), and Möller and Möller (1966).

Various hypotheses have been advanced to explain the phenomenon of immunological enhancement. They can be divided into two main categories: (1) antibodies act by changing various properties of the tumor cells, making them capable of growing in foreign hosts (Kaliss, 1958; Feldman and Globerson, 1960; Gorer, 1958); and (2) antibodies induce enhancement by interfering with the homograft reaction of the host (Snell, 1956; Billingham et al., 1956).

A number of experiments have made the first group of hypotheses unlikely. Thus, it has not been possible to demonstrate that antibodies directly change different tumor cell properties, such as growth rate, homotransplantability, and rate of synthesis of cellular isoantigens (Möller, 1963b). Furthermore, the enhancement phenomenon shows a stringent requirement for immunological specificity and can be induced only with antibodies produced in the tumor-recipient strain. Tumor enhancement cannot be achieved with the same antibodies in an immunologically different host nor in the same host with antibodies produced in a different strain against the same tumor (Möller, 1963a). These findings argue against a direct effect of antibodies on the tumor cells. The relevant variable for the induction of enhancement appears to be the presence of antibodies against *all* major foreign antigens on the tumor cells. It has been demonstrated by various authors (Snell *et al.*, 1960; Brent and Medawar, 1961; Möller, 1963a,c) that pretreatment of animals with antibodies directed against foreign tumor homografts efficiently suppresses the development of cell-mediated immunity detected by different methods. Antibody pretreatment also suppresses the development of humoral isoantibodies against the tumor homograft (Möller, 1963a,c). Suppression is highly specific and was only observed with regard to those histocompatibility antigens against which antibodies were present. Other antigens present on the same tumor cell stimulated a normal immune response (Möller, 1963c).

The level at which humoral antibodies exert their suppressing effect on the immune response of the host has been the subject of much attention. A distinction has been made between three possible levels at which antibodies may interfere with the homograft reaction: the afferent, central, and efferent levels, respectively. Afferent suppression means that antibodies suppress the induction of the immune response by interfering with antigen stimulation by the foreign histocompatibility antigens on the tumor. Central suppression implies a direct effect of antibodies on the immunologically competent cells of the host. Efferent inhibition postulates that humoral antibodies protect the neoplastic cells from destruction by immune lymphoid cells of host origin, presumably by combining with the target tumor cell antigens. Since the afferent and efferent mechanisms both imply a reaction between antibodies and the transplanted cells, in contrast to the central mechanism, it is important to study whether antibodies cause enhancement by reacting with the target cells or act at the host level.

Several findings have suggested that enhancement is caused by an antibody reaction at the target tumor cell level. Thus, target tumor cells coated *in vitro* with humoral antibodies and subsequently washed carefully grow progressively in foreign untreated recipients (Möller,

1963a, 1964a). The possibility that the antibodies dissociated from the tumor cells and acted centrally in the hosts was made less likely by the following experiment (Möller, 1964a). Antibody-coated tumor cells were inoculated into one side and uncoated cells onto the other side of the same incompatible untreated host. Only the antibody-coated cells were enhanced. If antibodies had dissociated and acted centrally on the lymphoid cells of the host, enhancement would have been expected to occur also with the nonantibody-treated tumor, which was not the case.

Although several findings strongly support the concept that enhancement depends, to a large extent, on an afferent suppression of immunization, this reaction by itself cannot fully explain the enhancement phenomenon, since enhancement of tumor grafts may be achieved in presensitized hosts (Kaliss, 1958; Feldman and Globerson, 1960; Haskova and Svoboda, 1962). This finding is best explained by an efferent mechanism, i.e., protection of the tumor cells from destruction by immune lymphoid cells. In the test system described above (Möller, 1964a), the existence of such a mechanism is revealed, i.e., the uncoated tumor cells were capable of immunizing the recipients sufficiently to cause their own rejection, but in spite of this, the antibody-coated tumor cells grew progressively.

Direct attempts to demonstrate an antagonistic relationship between cell-mediated immune reactions and humoral antibodies have been performed both in vitro and in vivo (Batchelor and Silverman, 1962; Möller, 1963c; E. Möller, 1965b). In vivo, it has been demonstrated that tumor cells treated with humoral antibodies are partially or completely protected from the destructive influences exerted by admixed immune lymphoid cells (Möller, 1963c). However, in certain experimental conditions, synergistic effects between the two types of immune responses may be found (Batchelor and Silverman, 1962). This may be related in part to quantitative factors concerning the amount of antibody and the number of admixed lymphoid cells.

Although the in vivo findings by themselves strongly suggest the existence of an antagonistic effect between humoral and cell-mediated immune reactions, experiments of this type are complicated by the fact that immune lymphoid cells may also secrete humoral antibodies. Use of tissue culture systems provides a possible way of distinguishing between the two types of immune mechanisms. It has been clearly demonstrated by various authors (for refs. see Möller, 1965a) that cytotoxic effects in tissue culture may be obtained when immune lymphoid cells are added to allogeneic target cells. This effect is not complement-dependent, but requires close contact between the two cell populations. Furthermore, little effect is observed as a rule before 24 hours contact. In contrast, humoral antibodies are active within 1 hour and require the presence of complement. Thus, it is possible to study the antagonistic effect of serum antibodies on cell-mediated immunity by performing the experiments in the absence of complement. It has been shown (Möller, 1965b) that pretreatment of allogeneic target cells in tissue culture with humoral antibodies partly or completely suppressed the cytotoxic effect of admixed immune lymphoid cells. Presumably the antagonistic relationship between the two types of immune reactions depends upon competition for the same antigenic determinants on the target cells.

Thus, in summary, it is clearly established that immunological enhancement of tumor grafts depends basically upon the ability of antibodies to suppress the development of both the humoral and cell-mediated immune responses (afferent suppression). In addition, antibodies may also interfere with the effector mechanisms of the homo graft reaction and protect the graft from destruction by immune lymphoid cells (efferent suppression).

The enhancement phenomenon is not confined to antibody-resistant neoplastic cells transplanted across a strong histocompatibility barrier (H-2) but appears to be a general phenomenon in tissue transplantation. Thus, enhancement has been observed with normal tissues, such as skin (Billingham et al., 1956; Brent and Medawar, 1961) and ovaries (Möller, 1964b). It also exists with leukemic cells sensitive to cytotoxic isoantibodies (Boyse et al., 1962) and has been demonstrated in various non-H-2 systems (Möller, 1963b) and with tumor-specific antigens (Möller, 1964c). An analogous phenomenon has been observed after transfusion of H-2 incompatible mouse red cells into antibody- containing hosts. A proportion of such cells are not eliminated but change in sensitivity to hemolytic antibodies and become completely resistant (Möller, 1963d, 1964d). Since enhancement appears to be important for the progressive growth of incompatible neoplastic cells in immunologically competent hosts (Möller, 1964, 1965b,c), it may represent a mechanism by which autochthonous tumors possessing tumor-specific antigens may develop and grow progressively to the death of the hosts.

VIII. Relationship between Immunological Tolerance and Antibody-Mediated Suppression

Although immunological tolerance or paralysis and antibody-mediated suppression of the immune response are similar in so far as the expression of active antibody synthesis is suppressed, they appear to be different with regard to a number of other variables. It has been demonstrated in several recent experiments that induction of tolerance is influenced by the physical properties of the antigen. Thus, both HSA and BGG more efficiently induce tolerance if they are cleared of particulate matter by centrifugation (Dresser, 1962) or by passage through an animal (Frei et al., 1965). Addition of aggregated HSA to animal-passaged HSA decreases or abolishes the ability of the antigen to induce tolerance and increases its immunogenicity (Frei et al., 1965). Antibody-mediated suppression differs markedly from tolerance in this respect. It is achieved in many cases with aggregates of antigen and antibody (Uhr and Baumann, 1961a,b; Rowley and Fitch, 1964; Möller, 1963a; Möller and Wigzell, 1965), whereas tolerance induction usually fails after injection of antigen–antibody complexes. The duration of immunological tolerance appears to depend upon the persistence of the injected *antigenic* material. Antibody suppression, on the other hand, depends on the persistence of antibody (Uhr and Baumann, 1961a; Möller et al., 1967). Indeed, it has been shown with both toxoids and bacterial lipopolysaccharide antigens that the prolonged persistence of antigen results in reappearance of active antibody formation when the passively transferred antibody has disappeared or dissociated from the antigen (Uhr and Baumann, 1961a; Britton and Möller, 1967).

It is also generally accepted that immune lymphoid cells of tolerant animals have a central defect of their immunological capacity, and such cells do not immediately regain immunological competence if transferred to untreated syngeneic recipients (for discussion, see Medawar, 1960). In contrast, several experiments carried out with lymphoid cells from antibody-treated hosts demonstrated that the cells were fully competent to enter a normal primary immune response when transferred to an environment lacking detectable antibody (Möller, 1964a; McCullagh and Gowans, 1966; Wigzell, 1967).

Antibody-mediated suppression of the immune response may be a factor, however, in the development of immunological tolerance or paralysis as first suggested by Rowley and Fitch (1964, 1965a,b, 1967). They compared the number of PFC and the humoral antibody response to SRBC in rats given a single antigen injection at various times after birth and in rats repeatedly injected as newborns and adults. In both newborn and adults repeated injections of SRBC eventually caused a decline of the number of PFC, which subsequently stabilized at a low level, which was, however, considerably above the background level found in nontreated recipients. Thus, there was clear indication of an active immune response induced by the treatment, whereas the existence of tolerance was inferred from the low level of PFC and antibodies as compared to animals injected only once. It seems possible that the low number of PFC in repeatedly injected animals was caused by the difficulties in detecting 7 S PFC, which may have been present in large numbers, in contrast to the few 19 S-producing cells. Since no tests for 7 S PFC were carried out, this possibility cannot be excluded. The low titers of serum antibodies may also be ascribed to progressive changes in the class of antibody with continued immunization. Rowley and Fitch (1965a,b) suggested that active production of antibody in small quantities provides a homeostatic mechanism limiting the antibody response to subsequent antigen injection, i.e., antibody actively produced by relatively few antibody-forming cells is presumed to prevent stimulation of other potential antibody-forming cells that were not stimulated initially in the adult rats or that matured later in the growing tolerant rat. This homeostatic mechanism could account in part for the sustained suppression of the antibody response to SRBC produced in adult rats and for tolerance produced in newborn rats.

It follows that lymphoid cell populations from tolerant animals would contain a small proportion of antibody-forming cells, which could be detected by appropriate tests. However, this has not been the case with "high" antigen dose paralysis, and the failure to demonstrate specifically competent or committed cells forms the basis for assigning the phenomenon to a central failure of the immune system. The classic experiments using the discriminative graft-versus-host assay demonstrated the failure of specifically tolerant cells to mount an immune reaction against the inducing antigens (Simonsen, 1962b). Analogous conclusions have been obtained with noncellular antigens. However, it has been recently shown that tolerance and immunity can coexist (Mitchison, 1964; Nossal and Austin, 1966), particularly in induction of low antigen dose paralysis. Also, the suggestion that a small proportion of immunologically committed cells may suppress the induction of a primary immune response in noncommitted cells has been experimentally confirmed. Thus, it was demonstrated (Morris and Möller, 1967) that transfer of 7 S-producing cells into nonimmune syngeneic animals caused a marked suppression of the development of 19 S PFC subsequent to antigen injection, whereas a secondary response of 7 S PFC occurred. There is no indication, however, that immunity is a necessary prerequisite for the expression of tolerance. In certain systems the speed of induction of immunological tolerance (Simonsen, 1962a; Mitchison, 1964) suggests that it can be induced without antibody formation. In other cases immunity has been shown to precede paralysis (Mitchison, 1964; Dorner and

Uhr, 1964; Siskind and Howard, 1966). Whether immunity under certain conditions facilitates induction of tolerance remains an open question.

IX. Concluding Discussion

A. MECHANISMS OF SUPPRESSION

Although the preceding studies indicate that serum antibody plays a role in the regulation of antibody synthesis, the extent of this role is in doubt. This uncertainty results from the fact that virtually all studies of suppression have involved the injection of "extra" antibody to immunized animals, usually at a step of immunization when qualitatively similar antibody is not present in similar concentrations. One exception to this generalization is the study of Britton and Möller (1967) of the cyclical variations in the serum antibody and PFC response to a single injection of *Escherichia coli* lipopolysaccharide in mice. Their study suggests an important role for serum antibody in the physiological regulation of the immune response to this particular antigen. However, the other experiments discussed above are not physiological. Tentatively, it appears that serum antibody plays a major role in the regulation of antibody formation to nonmetabolizable antigens. Indeed, serum antibody may be the only means by which immunogen can be functionally inactivated in cases where metabolism through the usual apparatus of the reticuloendothelial system does not occur efficiently. With easily metabolizable protein antigens, serum antibody appears to be less important in the regulation of the antibody response. For example, moderate amounts of passive antibody specific to KLH or ϕX and obtained from primarily immunized animals are relatively ineffective in suppression (Finkelstein and Uhr, 1964; Dixon et al., 1967). Antibody capable of suppressing the primary response to these antigens develops late in immunization. It remains unclear, therefore, whether such antibody is effective in suppression in the immunized animals from which it was removed. These two antigens are unusually good immunogens, however, and with more conventional immunogens, antibody suppression may play a more significant role even in the early stages of immunization. The previously discussed findings that transfer of immune lymphoid cells from a primarily immunized animal to untreated syngeneic hosts resulted in suppression of the immune response of host cells subsequent to antigen (Morris and Möller, 1967), suggest a regulating role of antibody also in the primary response. Furthermore, in the human, passive antibody to Rh antigens can dramatically suppress the predicted formation of undesirable Rh antibodies in mothers already immunized through incompatible pregnancies (Clarke et al., 1963; Freda et al., 1964).

It is possible that serum antibody also plays a role in shifting the specificity of the antibody response to multiple antigenic stimulation as suggested by Brody *et al.* (1967). These authors have demonstrated that antigenic competition between two haptenic determinants located on separate protein carriers can be partially eliminated by the administration of antibody to one of the haptenic determinants. Thus, antibody-induced suppression may be a useful mechanism for dealing with antigenic competition arising during an infection. If the initial antibodies formed by an infected host should be ineffective in conferring protection, a mechanism to shift the immune response would be advantageous.

Although there is no universal agreement as to the mechanisms underlying suppression, the evidence cited previously leads us to conclude that the first step in suppression is the interaction of antibody with specific antigenic determinants. This conclusion is deduced primarily from the specificity of suppression. Moreover, the alternative possibility that parts of the antibody molecule other than the combining sites play a role is made less likely by the findings that $5\,S\,F(ab')_2$ and $3.5\,S$ Fab antibody fragments are moderately efficient in suppression. These observations argue against a direct interaction of specific antibody with any type of nucleic acid informational molecule or with the cell surface of a potential specific-antibody-forming cell, unless antigenic determinants are present on these structures. This latter possibility has been suggested by Rowley and Fitch (1967) who hypothesize that specifically competent lymphoid cells have antigenlike structures on their surfaces as well as specific antibody. Evidence discussed previously, however, suggests that the experimental basis for this hypothesis is consistent with other simpler explanations. The concept of interaction between suppressive antibody and antigen is in agreement with an increasing body of evidence suggesting that antigen can persist in the lymphoid tissues for considerable periods of time and continue to stimulate and influence the immune response as discussed above. Presumably, passive antibody interacts with such antigen and interrupts the continued stimulation of the immune system.

The concept that the first step in suppression is interaction between antibody and antigen destined to stimulate further immunization, facilitates the understanding of many otherwise puzzling observations of suppression. Thus, it is known that the average binding affinity of serum antibody increases during immunization (Eisen and Siskind, 1964). This increase presumably reflects analogous changes in the population of cells synthesizing antibody in which the receptors on such cells are similar to the product of the cell. It follows that antigen reacts preferentially with and, thus, selects cells having a high affinity receptor. These thermodynamic considerations allow understanding of suppression in the following terms. The injection of serum antibody into an animal to be immunized (or which has just been immunized) introduces a third partner in the interaction between antigen and potential antibody-forming cells. In the nonimmunized animal with a population of potential antibody-forming cells of low average affinity for antigen, introduced hyperimmune antiserum can effectively bind antigen and thus deprive the vast majority of potential antibody-forming cells of stimulation. In contrast, animals immunized many weeks previously possess a population of lymphoid cells that are ready to respond to a second injection of antigen ("memory" cells) and which have relatively high affinity antibody as receptors; therefore, the competition of serum antibody already present and even the introduction of higher concentrations of serum antibody is less effective in suppressing the secondary response. In other words, with increasing time after immunization, antibody molecules are formed with greater capacity to suppress, and at the same time, the immune mechanism becomes less suppressible. Therefore, the quantitative balance between the two factors will determine in large measure the extend (if any) of suppression.

The concept presented also allows an explanation of certain findings that might appear contradictory. Thus, different findings have been published concerning the possibility of suppressing 7 S antibody synthesis after it has been induced. An analogous discrepancy exists with regard to suppression of the secondary response. It seems likely that the success in suppressing 7 S synthesis was caused by the use of large amounts of antibodies obtained after hyperimmunization, as pointed out by the author (Wigzell, 1966a). The ability to suppress 19 S but not a continued 7 S response by transfer of 7 S-producing cells into untreated recipients subsequently injected with antigen (Morris and Möller, 1967), may also be explained in these terms: the 7 S antibodies produced by the transferred cells may efficiently compete with 19 S cells in the host for the antigen, whereas the antibodies cannot compete with the cells that previously synthesized these antibodies.

It is, as yet, not known at what site the interaction between suppressive antibody and antigen occurs. It seems established that this site need not be in the circulation, because antibody injected subsequent to clearing of the antigen from the circulation is still effective in preventing the expected antibody response. Although there is considerable evidence to suggest that antigen must be processed by the macrophages in order to act as immunogen (Fishman, 1961), the possibility also exists that antigen deposited directly on the surface of macrophages, reticulum cells, or lymphoid cells may be the relevant immunogen. There is no evidence at present for a decisive choice between these two possibilities and, therefore, for an answer to the site of interaction of immunogen and suppressive antibody. The possibility that antigen within antibody-forming cells is responsible for continued stimulation of these cells is not supported by the findings of Nossal et al. (1965a,b). They failed to find labeling in cells forming antibody to flagellin, although the efficiency of labeling with ^{125}I was claimed to be sufficient to allow detection of several molecule of antigen per cell; it cannot be assumed, however, that ^{125}I was attached to the antigenic determinants of the flagellin.

The mechanism by which antibody can reduce the immunogenicity of antigen after combining with it are not yet known. The simplest explanation is that the antigenic determinants on the antigen molecule are rendered sterically unavailable to the antibody-forming mechanism because of this binding. This possibility would suggest that escape from suppression would be possible if dissociation of antigen or digestion of antibody from antigen with resultant exposure of antigenic determinants occurred. In support of this possibility is the finding of dissociation of diphtheria toxin in vivo from complexes of diphtheria toxin–antitoxin that were formed under conditions of antitoxin excess (Uhr and Baumann, 1961a). Another possibility is that the metabolism of antigen may be altered after binding with antibody. This could result from a change in the rate of degradation of antigen or from an altered localization of antigen from an immunogenic to a nonimmunogenic site after combination with antibody. Further information about these possibilities depends to a large extent upon the ability to trace the relevant immunogen within lymphoid organs.

B. Practical Implications of Suppression

The principle of antibody suppression has been applied successfully in clinical medicine to prevent Rh hemolytic disease of the newborn. As discussed previously it was shown by Stern et al. (1956) that antibodies to A or B antigens prevented immunization to Rh antigens present on the same red cells. More relevant, however, was the finding that Rh-negative male volunteers injected with Rh-positive red cells coated with anti-D did not form anti-D (Stern et al., 1961). This specific suppression of immunity was used in the treatment of mothers expected to have Rh-incompatible fetuses by giving them injections of anti-D shortly after delivery (Finn et al., 1961; Freda and

Gorman, 1962). None of the treated mothers developed definite anti-D, whereas as many as 30% of untreated mothers became immunized as expected (Freda *et al.*, 1966; Clarke, 1966). The fact that antibody treatment after delivery was successful is related to the fact that in the majority of cases the major antigenic stimulation occurs during delivery (see Clarke, 1966). This procedure therefore is designed to prevent the development of Rh hemolytic disease of the fetus or newborn in future pregnancies.

Passive transfer of antibodies against histocompatibility antigens often results in prolonged survival of subsequent normal or neoplastic tissue grafts (enhancement). As yet this principle has not been adopted in clinical transplantation to ensure prolonged graft survival, but it seems possible that a practical application of this type may be achieved, provided that the outcome of the treatment can be predicted beforehand. Since suppression of the homograft reaction will most likely occur after antibody transfer, the critical variable which determines the outcome of the treatment is the sensitivity of the graft to humoral antibodies. It seems possible that an increased knowledge of the histocompatibility factors in humans and the development of test methods to study target cell sensitivity to serum antibodies in defined histocompatibility systems will make possible the practical application of the enhancement principle in human organ transplantation, at least in selected cases.

The successful suppression of allergic encephalomyelitis in experimental animals by transfer of serum antibody suggests the possible application of the principle in treatment of autoimmunity. As yet attempts in this direction have not been reported. Complications analogous to those encountered in transplantation systems are likely to occur, and it seems essential to establish target cell sensitivity to serum antibodies in each case before clinical trials are attempted.

Experimental evidence indicates that most, if not all, animal tumors possess tumor-specific antigens absent from normal cells of the autochthonous host. It seems likely that analogous findings will eventually be made in humans. Although the relationship between the specificity of the antigens in different tumors and their etiology is complex, the detection of tumor-specific antigenicity in humans opens a large area for the application of immunological techniques in cancer therapy and prophylaxis. Whether active or passive immunization will be used, the possible expression of immunological enhancement of the tumors exists. Actually it has been demonstrated that enhancement can be induced with passively transferred antibodies against tumor-specific antigens in several species and that active immunization with antigenic preparations derived from tumors containing tumor-specific antigens may lead to accelerated tumor development. As yet the outcome of an individual experiment of this type cannot be predicted. It follows that immunotherapy and prophylaxis of neoplasia cannot be applied to humans with success until the principles responsible for immunological enhancement are known and the result of an individual attempt can be predicted.

ACKNOWLEDGMENTS

The work by the authors presented here was supported in part by USPHS Grant No. AI-01821-10 and by the Commission on Immunization of the Armed Forces Epidemiological Board, in part by the Office of the Surgeon General, Department of the Army, Washington, and by the Swedish Medical Research Council and the Swedish Cancer Society. The authors wish to express their appreciation to Dr. Gregory Siskind, Dr. Erna Möller, and Dr. Hans Wigzell for helpful suggestions and criticisms of this review.

REFERENCES

Ada, G. L., and Williams, J. M. (1966). *Immunology* 10, 417.
Ada, G. L., Nossal, G. J. V., and Austin, C. M. (1964). *Australian J. Exptl. Biol. Med. Sci.* 42, 331.
Andersson, B., Wigzell, H., and Klein, G. (1967). *Transplantation* 5, 11.
Barr, M., Glenny, A. T., and Randall, K. J. (1950). *Lancet* 1, 6.
Batchelor, J. R., and Silverman, M. S. (1962). *Ciba Found. Symp. Transplantation* p. 216.
Bauer, D. C. (1963). *J. Immunol.* 91, 323.
Benacerraf, B., and Gell, P. G. H. (1959). *Immunology* 2, 53.
Benacerraf, B., and Gell, P. G. H. (1962). *In* "Mechanism of Cell and Tissue Damage Produced by Immune Reactions" (P. Grabar and P. Miescher, eds.), p. 136. Benno Schwabe, Basel.
Berg, O., and Källén, B. (1962). *Acta Pathol. Microbiol. Scand.* 54, 425.
Billingham, R. E., Brent, L., and Medawar, P. B. (1956). *Transplant. Bull.* 3, 84.
Boyden, S. (1964). *In* "Molecular and Cellular Basis of Antibody Formation" (J. Sterzl, ed.), p. 329. Czech. Acad. Sci., Prague.
Boyse, E. A., Old, L. J., and Stockert, E. (1962). *Nature* 194, 1142.
Brambell, F. W. R., Hemmings, W. A., Oakley, C. L., and Porter, R. R. (1960). *Proc. Roy. Soc. (London)* B151, 478.
Brent, L., and Medawar, P. B. (1961). *Proc. Roy. Soc. (London)* B155, 392.
Britton, S., and Celada, F. (1966). *Immunology*, in press.
Britton, S., and Möller, G. (1966). *In* "Genetic Variations in Somatic Cells," p. 213. Czech. Acad. Sci., Prague.
Britton, S., and Möller, G. (1967). To be published.
Britton, S., Wepsic, T., and Möller, G. (1967). *Immunology*, in press.
Brody, N. I., Walker, J. G., and Siskind, G. W. (1967). Unpublished observations.
Buxton, J. B., and Glenny, A. T. (1921). *Lancet* 2, 1109.
Campbell, D. H., and Garvey, J. S. (1963). *Advan. Immunol.* 3, 261, 37.
Celada, F., and Wigzell, H. (1966). *Immunology* 11, 453.

Clarke, C. A. (1966). *Vox Sanguinis* 11, 641.

Clarke, C. A., Donohoe, W. T. A., McConnell, R. B., Woodrow, J. C., Finn, R., Krevans, J. R., Kulke, W., Lehane, B., and Sheppard, P. M. (1963). *Brit. Med. J.* i, 979.

Cooper, M. D., Peterson, R. D. A., South, M. A., and Good, R. A. (1966). *J. Exptl. Med.* 123, 75.

Dixon, F. J. (1966). Personal communication.

Dixon, F. J., Talmage, D. W., and Maurer, P. H. (1952). *J. Immunol.* 68, 693.

Dixon, F. J., Jacot-Guillarmod, H., and McConahey, P. J. (1966). *J. Immunol.* 97, 350.

Dixon, F. J., Jacot-Guillarmod, H., and McConahey, P. J. (1967). *J. Exptl. Med.* 125, 1119.

Dorner, M. M., and Uhr, J. W. (1964). *J. Exptl. Med.* 120, 435.

Downie, A. W., Glenny, A. T., Parish, H. J., Spooner, E. T. C., Vollum, R. L., and Wilson, G. S. (1948). *J. Hyg.* 46, 34.

Dray, S. (1962). *Nature* 195, 677.

Dresser, D. W. (1962). *Immunology* 5, 378.

Dresser, D. W. (1967). Personal communication.

Dresser, D. W., and Wortis, H. H. (1965). *Nature* 208, 859.

Dubiski, S., and Fradette, K. (1966). *Proc. Soc. Exptl. Biol. Med.* 122, 126.

Eisen, H. N., and Siskind, G. W. (1964). *Biochemistry* 3, 996.

Erickson, R. P., Herzenberg, L. A., and Goor, R. (1964). *Transplantation* 2, 175.

Fahey, J. L., and Robinson, A. G. (1963). *J. Exptl. Med.* 118, 845.

Feldman, M., and Globerson, A. (1960). *J. Natl. Cancer Inst.* 25, 631.

Finkelstein, M. S., and Uhr, J. W. (1964). *Science* 146, 67.

Finkelstein, M. S., and Uhr, J. W. (1966). *J. Immunol.* 97, 565.

Finn, R., Clarke, A., Donohoe, W. T. A., McConnell, R. B., Sheppard, P. M., Lehane, B., and Kulke, W. (1961). *Brit. Med. J.* i, 1486.

Fishman, M. (1961). *J. Exptl. Med.* 114, 837.

Fitch, F. W., and Rowley, D. A. (1967). Unpublished observations.

Franklin, E. C. (1964). *Progr. Allergy* 8, 58.

Franklin, E. C., and Ovary, Z. (1963). *Immunology* 6, 434.

Franzl, R. E. (1962). *Nature* 195, 457.

Franzl, R. E., and Marello, J. (1966). Personal communication.

Freda, V. J., and Gorman, J. G. (1962). *Bull. Sloane Hosp. Women* 8, 147.

Freda, V. J., Gorman, J. G., and Pollack, W. (1964). *Transfusion* 4, 26.

Freda, V. J., Gorman, J. G., and Pollack, W. (1966). *Science* 151, 828.

Frei, P. C., Benacerraf, B., and Thorbecke, G. J. (1965). *Proc. Natl. Acad. Sci. U.S.* 53, 20.

Garvey, J. S., and Campbell, D. H. (1956). *J. Immunol.* 76, 36.

Gell, P. G. H., and Benacerraf, B. (1959). *Immunology* 2, 64.

Glenny, A. T., and Sudmersen, H. J. (1921). *J. Hyg.* 20, 176.

Gorer, P. A. (1958). *Ann. N.Y. Acad. Sci.* 73, 707.

Gorer, P. A., and Kaliss, N. (1959). *Cancer Res.* 19, 824.

Haskova, V., and Svoboda, J. (1962). *Mech. Immunol. Tolerance, Proc. Symp., Liblice, Czech., 1961.* p. 431.

Hellström, K.-E., and Möller, G. (1965). *Progr. Allergy* 9, 158.

Herzenberg, L. A., and Herzenberg, L. A. (1966). *In* "Genetic Variation in Somatic Cells," p. 227. Czech. Acad. Sci., Prague.

Horibata, K., and Uhr, J. W. (1967). *J. Immunol.* **98**, 972.

Humphrey, J. H. (1960). *Mech. Antibody Formation, Proc. Symp., Prague, 1959* pp. 39–43.

Humphrey, J. H., and Fahey, J. L. (1961). *J. Clin. Invest.* **40**, 1696.

Ingraham, J. S., and Bussard, A. (1964). *J. Exptl. Med.* **119**, 667.

Ishizaka, K., Ishizaka, T., and Hornbrook, M. M. (1966). *J. Immunol.* **97**, 840.

Jerne, N. K. (1965). *In* "Molecular and Cellular Basis of Antibody Formation" (J. Sterzl, ed.), p. 459. Czech. Acad. Sci., Prague.

Jerne, N. K. (1966). Personal communication.

Jerne, N. K., and Nordin, A. A. (1963). *Science* **140**, 405.

Jerne, N. K., Nordin, A. A., and Henry, C. (1963). *In* "Cell-bound Antibodies" (B. Amos and H. Koprowski, eds.), p. 109. Wistar Inst. Press, Philadelphia, Pennsylvania. p. 109.

Kaliss, N. (1958). *Cancer Res.* **18**, 992.

Kaliss, N. (1962). *Ann. N.Y. Acad. Sci.* **101**, 64.

Kalmanson, G. M., and Bronfenbrenner, J. (1943). *J. Immunol.* **47**, 387.

Karush, F., and Eisen, H. N. (1962). *Science* **136**, 1032.

Koprowski, H., and Fernandes, M. W. (1962). *J. Exptl. Med.* **116**, 467.

Kennedy, J. C., Till, J. E., Siminovitch, L., and McCulloch, E. A. (1966). *J. Immunol.* **96**, 973.

Landy, M., Sanderson, R. P., Bernstein, M. T., and Lerner, E. M. (1965). *Science* **147**, 1591.

Levine, B. B. (1965). *J. Exptl. Med.* **121**, 873.

Levine, P. (1943). *J. Heredity* **34**, 71.

Locke, R. F., Segre, D., and Myers, W. L. (1964). *J. Immunol.* **93**, 576.

McBride, R. A., and Schierman, L. W. (1966). *Science* **154**, 655.

McConahey, P. J., and Dixon, F. J. (1967). *Federation Proc.* **26**, 700.

McCullagh, P. J., and Gowans, J. L. (1966). Personal communications.

Mäkelä, O., and Nossal, G. J. V. (1961). *J. Immunol.* **87**, 447.

Medawar, P. B. (1960). *Ciba Found. Symp. Cellular Aspects Immunity* p. 134.

Merchant, B., and Hraba, T. (1966). *Science* **152**, 1378.

Miller, J. J., and Nossal, G. J. V. (1965). *J. Exptl. Med.* **120**, 1075.

Mitchell, J., and Abbot, A. (1965). *Nature* **208**, 500.

Mitchison, N. A. (1964). *Proc. Roy. Soc. (London)* **B161**, 275.

Mitchison, N. A. (1965). *Immunology* **9**, 129.

Möller, E. (1964). *J. Natl. Cancer Inst.* **33**, 979.

Möller, E. (1965a). *Science* **147**, 873.

Möller, E. (1965b). *J. Exptl. Med.* **122**, 11.

Möller, E. (1965c). *J. Natl. Cancer Inst.* **35**, 1053.

Möller, E., and Möller, G. (1962). *J. Exptl. Med.* **115**, 537.

Möller, E., Britton, S., and Möller, G. (1967). *In* "Regulation of the Antibody Response" (B. Cinader, ed.). In press.

Möller, G. (1963a). *J. Natl. Cancer Inst.* **30**, 1153.

Möller, G. (1963b). *J. Natl. Cancer Inst.* **30**, 1177.

Möller, G. (1963c). *J. Natl. Cancer Inst.* **30**, 1205.

Möller, G. (1963d). *Nature* **199**, 573.

Möller, G. (1964a). *Transplantation* **2**, 405.

Möller, G. (1964b). *Transplantation* **2**, 281.

Möller, G. (1964c). *Nature* **204**, 846.

Möller, G. (1964d). *Nature* **202**, 357.
Möller, G. (1965). *Nature* **207**, 1166.
Möller, G. (1966). *J. Immunol.* **96**, 430.
Möller, G., and Möller, E. (1966). *In* "Antibodies to Biologically Active Molecules" (B. Cinader, ed.), p. 349. Macmillan (Pergamon), New York.
Möller, G., and Wigzell, H. (1965). *J. Exptl. Med.* **121**, 969.
Mollison, P. L. (1951). "Blood Transfusion in Clinical Medicine." Blackwell, Oxford.
Morris, A., and Möller, G. (1967). To be published.
Nagano, Y., and Takeuti, S. (1951). *Japan J. Exptl. Med.* **21**, 427.
Nakano, M., and Braun, W. (1966). *Science* **151**, 338.
Nikanoroff, P. J. (1897). *Arch. Sci. Biol. St. Petersburg* **6**, 57.
Nossal, G. J. V. (1966). *Ann. N.Y. Acad. Sci.* **129**, 822.
Nossal, G. J. V., and Austin, C. M. (1966). *Australian J. Exptl. Biol. Med. Sci.* **44**, 327.
Nossal, G. J. V., Ada, G. L., and Austin, C. M. (1964). *Australian J. Exptl. Biol. Med. Sci.* **42**, 311.
Nossal, G. J. V., Ada, G. L., and Austin, C. M. (1965a). *J. Exptl. Med.* **121**, 945.
Nossal, G. J. V., Ada, G. L., Austin, C. M., and Pye, J. (1965b). *Immunology* **9**, 349.
Olhagen, B., Birke, G., Plantin, L. O., and Ahlinder, S. (1963). *Protides Biol. Fluids, Proc. Colloq.* **11**.
Olitzki, L. (1953). *J. Immunol.* **29**, 453.
Otten, L., and Hennemann, J. P. (1939). *J. Pathol. Bacteriol.* **49**, 213.
Papermaster, B. W. (1967). Personal communication.
Pappenheimer, A. M., Jr., Lundgren, H. P., and Williams, J. W. (1940). *J. Exptl. Med.* **71**, 247.
Park, W. H. (1922). *J. Am. Med. Assoc.* **79**, 1584.
Paterson, P. Y. (1960). *J. Exptl. Med.* **111**, 119.
Paterson, P. Y. (1966). *Advan. Immunol.* **5**, 130.
Paterson, P. Y., and Harwin, S. M. (1963). *J. Exptl. Med.* **117**, 755.
Pearlman, D. S. (1966). *Federation Proc.* **25**, 548.
Playfair, J. H. L., Papermaster, B. W., and Cole, L. J. (1965). *Science* **149**, 998.
Pokorna, Z., and Vojtiskova, M. (1966). *Folia Biol. (Prague)* **12**, 88.
Ramon, G., and Zoeller, C. (1933). *Compt. Rend. Soc. Biol.* **112**, 347.
Regamey, R. H., and Aegerter, W. (1951). *Schweiz. Z. Allgem. Pathol. Bakteriol.* **14**, 554.
Rowe, D. S., and Fahey, J. L. (1965). *J. Exptl. Med.* **121**, 185.
Rowley, D. A., and Fitch, F. W. (1964). *J. Exptl. Med.* **120**, 987.
Rowley, D. A., and Fitch, F. W. (1965a). *J. Exptl. Med.* **121**, 671.
Rowley, D. A., and Fitch, F. W. (1965b). *J. Exptl. Med.* **121**, 683.
Rowley, D. A., and Fitch, F. W. (1967). *In* "Regulation of the Antibody Response" (B. Cinader, ed.). In press.
Sahiar, K., and Schwartz, R. S. (1964). *Science* **145**, 395.
Sahiar, K., and Schwartz, R. S. (1965). *J. Immunol.* **95**, 345.
Sahiar, K., and Schwartz, R. S. (1966). *Intern. Arch. Allergy Appl. Immunol.* **29**, 52.
Salvin, S. B., and Smith, R. F. (1960a). *J. Immunol.* **84**, 449.
Salvin, S. B., and Smith, R. F. (1960b). *J. Exptl. Med.* **111**, 465.
Segre, D., and Kaeberle, M. L. (1962). *J. Immunol.* **89**, 782.

Simonsen, M. (1962a). *Ciba Found. Symp. Transplantation* p. 185.
Simonsen, M. (1962b). *Progr. Allergy* **6**, 349.
Siskind, G. W., and Howard, J. G. (1966). *J. Exptl. Med.* **124**, 417.
Smith, R. T., and Eitzman, D. V. (1963). *Pediatrics* **33**, 163.
Smith, T. (1909). *J. Exptl. Med.* **11**, 241.
Snell, G. D. (1956). *Transplant. Bull.* **3**, 83.
Snell, G. D. (1963). *In* "Conceptual Advances in Immunology and Oncology," p. 323. Harper & Row (Hoeber), New York.
Snell, G. D., Winn, H. J., Stimpfling, J. H., and Parker, S. J. (1960). *J. Exptl. Med.* **112**, 293.
Spiegelberg, H. L., and Weigle, H. O. (1965). *J. Exptl. Med.* **121**, 323.
Stern, K., Davidsohn, I., and Masaitis, L. (1956). *Am. J. Clin. Pathol.* **26**, 833.
Stern, K., Goodman, H. S., and Berger, M. (1961). *J. Immunol.* **87**, 189.
Sterzl, J., and Mandel, L. (1964). *Folia Microbiol.* (*Prague*) **9**, 173.
Sterzl, J., and Riha, I. (1965). *Nature* **208**, 858.
Sterzl, J., Jilek, M., Vesely, J., and Mandel, L. (1966). *In* "Genetic Variation in Somatic Cells," p. 233. Czech. Acad. Sci., Prague.
Svehag, S. E., and Mandel, B. (1964). *J. Exptl. Med.* **119**, 21.
Tao, T. W., and Uhr, J. W. (1966). *Nature* **212**, 208.
Taranta, A., and Franklin, E. C. (1961). *Science* **134**, 1981.
Terres, G., and Stoner, R. D. (1962). *Proc. Soc. Exptl. Biol. Med.* **109**, 88.
Terres, G., and Wolins, W. (1959). *Proc. Soc. Exptl. Biol. Med.* **102**, 632.
Terres, G., and Wolins, W. (1961). *J. Immunol.* **86**, 361.
Uhr, J. W. (1964). *Science* **145**, 457.
Uhr, J. W., and Baumann, J. B. (1961a). *J. Exptl. Med.* **113**, 935.
Uhr, J. W., and Baumann, J. B. (1961b). *J. Exptl. Med.* **113**, 959.
Uhr, J. W., and Finkelstein, M. S. (1963). *J. Exptl. Med.* **117**, 457.
Uhr, J. W., and Finkelstein, M. S. (1967). *Progr. Allergy* **10**, 37.
Uhr, J. W., and Pappenheimer, A. M., Jr. (1958). *J. Exptl. Med.* **108**, 891.
Uhr, J. W., Salvin, S. B., and Pappenheimer, A. M., Jr. (1957). *J. Exptl. Med.* **105**, 11.
Van Behring, and Wernicke (1892). *Z. Hyg. Infektionskrankh* **12**, 10.
Walker, J. G., and Siskind, G. W. (1967). Unpublished observations.
Warner, N. L., Szenberg, A., and Burnet, F. M. (1962). *Australian J. Exptl. Biol. Med. Sci.* **40**, 373.
Weiler, E., Melletz, E. W., and Breuninger-Peck, E. (1965). *Proc. Natl. Acad. Sci. U.S.* **54**, 1310.
Wigzell, H. (1966a). *J. Exptl. Med.* **124**, 953.
Wigzell, H. (1966c). *J. Immunol.* **97**, 608.
Wigzell, H. (1967). Ph.D. Thesis, Balders Tryckeri, Stockholm.
Wigzell, H., Möller, G., and Andersson, B. (1966). *Acta Pathol. Microbiol. Scand.* **66**, 530.
Winn, H. J. (1959). *Natl. Cancer Inst. Monograph* **2**, 113.
Winn, H. J. (1962). *Ann. N.Y. Acad. Sci.* **101**, 23.
Winn, H. J. (1965). *Ciba Found. Symp. Complement* p. 133.

The Mechanism of Immunological Paralysis

D. W. DRESSER AND N. A. MITCHISON

National Institute for Medical Research, London, England

I. Introduction

The subject of immunological paralysis derives its principal practical interest from possible future applications in surgery. The barrier that the homograft reaction imposes to tissue transplantation can be overcome by paralyzing the host with donor antigens. This form of treatment, which is already a routine procedure in animals, provides the best hope of suppressing a homograft reaction without interfering with other immunological defenses. Our understanding of paralysis also throws light on the workings of the immunological mechanism. The development of responsiveness, the mode of action of radiation and immunosuppressive drugs, the functions of macrophages and lymphocytes, and cellular recognition of antigens, are all topics on which paralysis has a bearing. The question whether paralysis is responsible for the failure of self-antigens to provoke a response still requires a final answer.

This review does not cover all these matters. It concentrates on the information that has been gained with noncellular, well-defined antigens and is largely confined to material published since the last review of the subject in this series by Smith (1961). Attention therefore focuses on protein antigens, particularly heterologous serum proteins. These are of special interest at present because of their application in exploring the induction of paralysis in adult animals by low doses. A note of uncertainty enters the discussion when theories based on this restricted material are extended to cellular and other antigens.

The arrangement of this review is in part historical. Treatment of the immature animal (Section II,A) belongs logically with other procedures for reducing concomitant immunization (Section II,D) but is dealt with separately because it is the way in which immunological tolerance was discovered and it subsequently attracted special attention. The distinction in name between "tolerance" and "paralysis" dates from the time when this similarity of effect was not properly understood. In view of the similarity that has now been established the distinction in name has been withdrawn, and the two terms can be used interchangeably. The cellular basis of paralysis (Section VI) is now sufficiently well understood to have become very much part of the definition; other types of antigen-specific immunosuppression involving antibody, immune deviation, or enhancement, are distinguished in Sections II,B, III, and VI. We do not attempt a formal definition until the summary.

The following abbreviations are used:

Haptens
    DNP   2,4-dinitrophenyl
    DNCB   2,4-dinitrochlorobenzene
    Picryl chloride   2,4,6-trinitrochlorobenzene
    NIP   4-hydroxy-3-iodo-5-nitrophenylacetic acid
    NDMA   3-nitrosodimethylaniline
Proteins
    G   serum globulins
    GG γ-globulins prepared by ethanol or salt fractionation
    SA   serum albumin

    B = bovine; C = chicken; Ca = canine; E = equine; H = human; L = leporine; P = porcine; R = rabbit; T = turkey

    γA, γG, γM   standard nomenclature for human γ-globulin classes (Cohen, 1965). (This nomenclature is used by us to name analogous proteins in other species.)

II. Conditions for the Induction of Paralysis

The circumstances under which paralysis can be induced are outlined and classified in this section. The degree of concomitant immunization normally decides whether paralysis will be induced by exposure to antigen, because immunization can both mask paralysis and also probably impede its induction (Section II,C). For this reason the discussion concentrates on circumstances that favor or prevent immunization, i.e., on immunogenic capacity and sensitivity to immunization.

A guide to work published since Smith's review (1961) on induction of paralysis is given in Table I. Although this table is not comprehensive, it does illustrate the diversity of animals and antigens that have been used. In most instances we attempt to cite the most recent references, when a particular system has been the subject of extensive work.

A. IMMATURITY OF THE ANIMAL

The pioneering work on immunological paralysis suggested that exposure to antigen at an early stage of development was required for induction (Burnet and Fenner, 1949; Owen, 1945; Billingham et al., 1953, 1956; Hašek, 1953; Hanan and Oyama, 1954; Cinader and Dubert, 1955). Accordingly, a "critical period" of susceptibility was postulated, during which paralysis but not immunity could be induced. However, in the past few years young animals, and sometimes even embryos, have been found capable of weak immune responses. For instance, Howard and Michie (1962) showed that transplantation immunity can be induced in newborn mice with smaller doses of allogeneic cells than are required to paralyze; and Silverstein et al. (1963) showed that fetal sheep can respond after 70 days of a 150-day gestation to bacteriophage, ferritin, and ovalbumin. The concept of a period of unique susceptibility to paralysis must, therefore, be abandoned. The induction of paralysis in adults will be dealt with in the remaining part of this section, but it is relevant here to quote the demonstration that transplantation tolerance can be induced during adult life, since the critical period was first postulated for the homograft reaction (Mariani et al., 1959; Brent and Gowland, 1962; Good et al., 1966).

No systematic difference can be found between the dose of antigen required to paralyze the newborn and that required for adults, provided that the antigen can be administered under nonimmunizing conditions

TABLE I

SELECTED KEY REFERENCES TO WORK ON IMMUNOLOGICAL PARALYSIS PUBLISHED
SUBSEQUENT TO THE REVIEWS OF SMITH (1961) AND HAŠEK et al. (1961)

Reference	Paralytogen; animal; other information
	VIRUS
Rubin et al. (1962)	Avian leucosis; chickens; infection
Schwartz et al. (1964)	Vaccinia, 'flu.; guinea pig; —
Seamer (1965)	LCM; mouse; infection
	BACTERIA
Gowland et al. (1965)	Pseudomonas NCMB 406; rabbit; —
	CARBOHYDRATE
Bathson et al. (1963)	Klebsiella polysaccharide; mouse; adults
Friedman (1964, 1966)	Shigella polysaccharide; mouse; —
Brooke (1965, 1966)	Pneumococcus polysaccharide; mouse; adult
Britton and Möller (1966)	Escherichia coli lipopolysaccharide; mouse; plaque assay
Gaines et al. (1966)	E. coli Vi; mouse; tested with Salmonella typhosa Vi
Siskind and Howard (1966)	Pseudomonas polysaccharide; mouse; —
	PROTEIN
Humphrey and Turk (1961); Turk and Humphrey (1961)	BSA, HGG; guinea pig; circulating Ab and hypersensitivity
Battisto and Miller (1962)	BGG; guinea pig; low-zone, injection via hepatic portal system
Dresser (1926b, 1965)	BGG; mouse; low-zone and high-zone
Weiss and Main (1962)	Diphtheria toxoid; guinea pig; —
Dietrich and Weigle (1963)	Var. SA and var. GG; mouse; —
Schwartz and Dameshek (1963)	BSA; rabbit; drugs, adults, etc.
Sercarz and Coons (1963)	BSA; mouse; single cells by immunofluorescence
Dorner and Uhr (1964)	BSA; rabbit; —
Humphrey (1964a,b)	BSA, HSA; rabbit; —
Ivanyi et al. (1964)	HSA; chicken; —
Mitchison (1964)	BSA; mouse; low-zone and high-zone
Schechter et al. (1964b)	HSA + peptide; rabbit; —
Dvorak et al. (1965, 1966)	HSA, BGG; rat, guinea pig; —
Frei et al. (1965)	BSA; rabbit; adult, screened antigen
Linscott and Weigle (1965a,b)	BSA; rabbit; cross-reaction and loss
Muller-Eberhard et al. (1965)	Human transferrin; rabbit; cross-reaction and loss
Nossal et al. (1965)	Salmonella flagellin; rat; —
Tempelis (1965)	Turkey GG, goose GG; chicken; —
Austin and Nossal (1966)	Salmonella flagellin; rat; —
Azar (1966)	HGG; mouse; —
Claman and Bronsky (1966)	BGG; mouse; high-zone and drugs
Cruchaud (1966)	BSA; rat; —

TABLE I (*Continued*)

Reference	Paralytogen; animal; other information
Day and Farr (1966)	BSA; rabbit; high-zone, low-binding Ab
Gell and Kelus (1966)	Allotype; rabbit; negative result
Phillips (1966)	Turkey GG, chicken; —
Staples *et al.* (1966)	BGG; rat; —
Streilein and Hildreth (1966)	BGG; guinea pig; —
Weigle (1966a,b)	BSA, HSA; mouse, rabbit, guinea pig; cross-reaction, loss

SYNTHETIC POLYPEPTIDES

Maurer *et al.* (1963)	Var.; rabbit; —
Sela *et al.* (1963)	Var; rabbit; —
Schechter *et al.* (1964a)	Var.; rabbit; —
Gill (1965)	Var.; rabbit; —
Janeway and Sela (1967)	L-, D-Polymers; mouse; —

HAPTEN

Battisto and Chase (1963, 1965)	Picryl, etc.; guinea pig; feeding
Coe and Salvin (1963)	DNCB; guinea pig; feeding
Frey *et al.* (1964)	Neoarsphenamine; guinea pig; intravenously
Lowney (1965)	DNCB; guinea pig; after topical application
Jones and Leskowitz (1965)	Arsenate–amino acid polymer; guinea pig; —
Leskowitz (1967)	Arsenate–tyrosine; guinea pig; adults

ERYTHROCYTES

Mitchison (1962a,b)	Allogeneic RBC; chicken; —
Rowley and Fitch (1965a,b)	Sheep RBC; rat; plaque assay
Thompson *et al.* (1966)	Allogeneic RBC; mouse; —

VARIOUS TRANSPLANTATION ANTIGENS

McKhann (1962, 1964)	H-1, H-2, H-3; mouse; adults
Medawar (1963)	Allogeneic tissue extracts; mouse; —
Anderson and Benirschke (1964)	Allogeneic by parity; armadillo; monozygotic quadruplets
Porter and Breyere (1964)	Allogeneic tissue; mouse; parity
McBride and Simonsen (1965)	Allogeneic tissue; mouse; parabiosis
Good *et al.* (1966)	Allogeneic tissue extracts; mouse; —
Bainbridge and Gowland (1966b)	Allogeneic tissue; mouse; low-zone and high-zone
Eichwald (1963)	(Review)
Gowland (1965)	(Review)

GENETIC UNRESPONSIVENESS

Sobey *et al.* (1966)	BSA; mouse; —

or in a nonimmunogenic form. This point has been examined in detail for BSA and lysozyme injected into mice (Mitchison, 1967, and unpublished data). Establishing nonimmunizing conditions in adults usually entails the use of radiation or immunosuppressive drugs, and it has been argued that animals treated in this way come to resemble the newborn in respect of cell number or cell type in lymphoid tissue (Schwartz and Dameshek, 1963). This treatment is not, however, necessary for weakly or nonimmunogenic antigens (as here defined), so it is useful to have, in BSA, an example of one such antigen for which the paralysis threshold of the newborn is known.

Nevertheless, the fact is well established that the newborn can usually be paralyzed more easily and more conveniently than the adult (for references, see Smith, 1961; and Table I). On the whole this can be ascribed simply to the well-known *relative* lack of responsiveness in young animals (Freund, 1930) that is presumably the result of deficiency of immunologically competent cells.

The work of Mitchell and Nossal (1966) raises the question whether the special immunological susceptibility of the young animal can be attributed at all to a failure of macrophage function. *Salmonella* flagellin labeled with ^{125}I was injected into newborn rats and was found to take up a wide distribution without undergoing normal phagocytosis. This might encourage paralysis, which the antigen under these conditions is known to induce (Nossal *et al.*, 1965). In interpreting these observations the question is whether the failure to localize flagellin prevented immunization, or, alternatively, whether lack of antibody prevented normal localization. In spite of the failure of passively administered antibody to bring about normal localization, the latter explanation seems the more probable in the light of the information now available concerning the role of antibody in parafollicular localization of antigen (discussed in Section VI). In that case the work provides an elegant demonstration of one way in which the lack of an immune response could facilitate the induction of paralysis.

Provision of additional competent cells by adoptive transfer impedes the induction of paralysis in newborn mice (Cohen and Thorbecke, 1963; Friedman, 1965c). The effect was originally ascribed to a reduction in the ratio of antigen to cells and was, therefore, interpreted as lending support to the hypothesis that a high dose of antigen *per cell* caused paralysis. It can equally well be interpreted according to the hypothesis advanced here, as increasing responsiveness to a level where the production of antibody masks and interferes with paralysis.

B. Immunogenic Capacity

1. Nonimmunogenic Antigens

A classification of the modes of response to a potential antigen is offered in Table II and Fig. 1. It includes the important category of nonimmunogenic antigens, introduced by Dresser (1961a): these antigens paralyze, but do not immunize unless administered in conjunction with adjuvant. The principal example of this category is one of the components of BGG, which has been extensively studied (Dresser, 1961b, 1962a,b, 1963, 1965, and unpublished; Claman, 1963; Dietrich and Weigle 1964). The protein in question is apparently γG, as judged by electro-

TABLE II[a,b]

A Potential Antigen May
1. Lack immunogenic capacity and
   (a) possess *antigenicity* (leading to antibody production only if an extrinsic adjuvant is present), or
   (b) lack *antigenicity*, due to
       (i) natural or acquired unresponsiveness (paralysis, tolerance) or
       (ii) genetic inability of the animal to synthesize certain antibody specificities.
2. Possess immunogenic capacity and
   (a) possess *antigenicity* (the conventional antigens), or
   (b) lack *antigenicity*, due to
       (i) natural or acquired unresponsiveness or
       (ii) genetic inability.[b]

[a] Immunity to autoantigens might ensue due to failure at 1(a) by a gain in immunogenic capacity (Dresser, 1961b), or at 2(b) by gaining antigenicity (Weigle, 1964c).

[b] "Natural unresponsiveness" and a genetic inability to respond may not be mutually exclusive mechanisms.

phoretic and chromatographic behavior, carbohydrate content, and blood clearance. This material elicits an immune response in CBA mice if injected with adjuvant, but otherwise induces paralysis in the sense that treated mice subsequently do not respond to challenge with the same protein injected with adjuvant. Another BGG component, γA, immunizes without requiring adjuvant. The presence of this immunogenic component in a mixture is not of itself sufficient to provoke a response against the γG component. The preparation of γG in a nonimmunogenic form is further discussed in Section II,B,3.

A comparison is made in Table III between the quantities needed to immunize and to paralyze, for γ-globulins and pneumococcal polysaccharides. For the nonimmunogenic γ-globulins, paralysis can be induced by approximately the same threshold dose as is required for immunization in the presence of adjuvant. Paralysis in this dose range can, therefore, hardly be ascribed to antigen excess.

The response to ethanol-fractionated BGG has been examined, with consistent results, by immune clearance, precipitation in gel, and passive hemagglutination. Fresh BγG is equally nonimmunogenic. Other strains

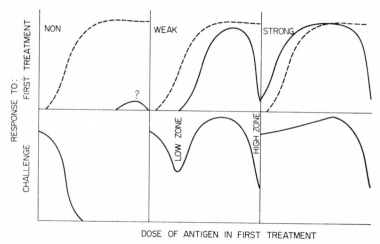

DOSE OF ANTIGEN IN FIRST TREATMENT

Fig. 1. A summary of induction of paralysis by different kinds of protein antigens. The antigens are classified as non, weak, and strong immunogens. In each graph the paralysis response (dashed line) and the immune response (solid line) increase upward and the dose of antigen used for the first treatment increases from left to right. The first treatment is by protein antigen in saline, and the challenge is by a standard dose of the same antigen in adjuvant.

of mice besides CBA manifest the same lack of response, including C57BL/6, A, and BALB/C. The major components of γ-globulin obtained from other species are also nonimmunogenic in CBA mice (Table IV). Outside the mouse, BGG is nonimmunogenic and paralyzes in guinea pigs, and the same is true of LGG in rabbits (Table IV).

Another example of a nonimmunogenic antigen comes from the work of Janeway and Sela (1967) on poly-D-amino acids in mice. These polypeptides provoke a response over a narrow dose range with a maximum at 1 μg., if injected in Freund's adjuvant, contrary to the conclusion of Zubay (1963). The mice are paralyzed by fairly low doses (0.1–1000 μg.) of D-polypeptide without adjuvant. Since larger doses of the corresponding poly-L-amino acid immunize without the requirement for

TABLE III

Immunological Paralysis Induced in Adult Inbred Mice and Outbred Himalayan Rabbits by Single Doses of Antigen[a]

(A) Animals	(B) Antigens	(C) Mol. wt. of whole molecule	(D) Mol. wt. of smallest determinant units[b]	(E) Immunogenicity when paralyzing	(F) Minimum paralyzing dose (gm. and moles)[c]	(G) Minimum No. of determinant units to paralyze mouse	(H) Minimum immunizing dose (gm. and moles)[c]	(I) Minimum No. of determinant units to immunize mouse	(J) Ratio F/H	(K) Paralysis "zone"
C3H	SSS I	170,000	180	+	5×10^{-4} / 2×10^{-6}	2×10^{18}	1×10^{-8} / (5×10^{-11})	5×10^{13}	5×10^{4}	High
C3H	SSS II	500,000	180	+	5×10^{-5} / 2×10^{-7}	2×10^{17}	3×10^{-9} / 2×10^{-11}	1×10^{15}	2×10^{4}	
C3H	SSS III	170,000	180	+	5×10^{-5} / 2×10^{-7}	2×10^{17}	1×10^{-7} / (5×10^{-10})	5×10^{14}	5×10^{2}	
CBA	Bovine γA	170,000	50,000	++	$>1 \times 10^{-4}$ / $>2 \times 10^{-5}$	$>1 \times 10^{18}$	$<1 \times 10^{-5\,d}$ / $<2 \times 10^{-10}$	$<1 \times 10^{14}$	$>1 \times 10^{4}$	
CBA	Bovine γG	170,000	50,000	++ e	5×10^{-4} / (1×10^{-5})	5×10^{18}	1×10^{-5} / 2×10^{-10}	1×10^{14}	5×10^{4}	
Rabbit	EGG	170,000	50,000	++	$>1 \times 10^{-4\,g}$ / $>(2 \times 10^{-5})^{h}$	$>8 \times 10^{17}$	$<1 \times 10^{-3\,f}$ / $<(2 \times 10^{-8})$	$<8 \times 10^{15}$	$>>1 \times 10^{2}$	
CBA	Bovine γG	170,000	50,000	−	1×10^{-5} / (2×10^{-10})	1×10^{14}	1×10^{-5} / (2×10^{-10})	1×10^{14}	1	
Rabbit	LGG	170,000	50,000	−	$<1 \times 10^{-4\,f}$ / $<(2 \times 10^{-9})$	$<8 \times 10^{14}$	$<1 \times 10^{-3\,f}$ / $<(2 \times 10^{-8})$	$<8 \times 10^{15}$	0.1–1	Low

[a] All values in column H (minimum immunizing dose with or without adjuvants) are of the same order of magnitude. All values above the dashed line (immunogenic) in column F (minimum paralyzing dose) are of the same order of magnitude (2×10^{17}–5×10^{18}), but below the line (nonimmunogenic) the values are three orders of magnitude lower.

An easy comparison with BSA or *Salmonella adelaidae* flagellar and flagellin is not possible because in adults paralysis to these antigens was induced by courses of repeated injections. Concentration of antigen in the milieu of the immunologically competent cell might well be the criterion to consider.

Column J (ratio) shows that in a situation where the antigen is nonimmunogenic, the same relative numbers of molecules are required to paralyze as to immunize (with an adjuvant). This implies that the basic mechanisms of immunity and paralysis are closely analogous and that an empirical difference between paralysis and immunity in terms of dose of antigen required, will only exist when the antigen is immunogenic and initiates antibody production.

References—Felton *et al.* (1955a); Kabat and Meyer (1961); Dresser (1962b, 1963, 1965); Dresser and Gowland (1964).

[b] Column D, smallest *determinant units*, or the smallest repeating units of immunological significance. This is based on the smallest identical chemical entity which could be haptenic or antigenic; for instance, sugars in carbohydrates or peptide chains in proteins.

[c] In presence of an adjuvant if antigen normally nonimmunogenic.

[d] SSS I is Type I pneumococcal polysaccharide. (See also Section I.)

[e] γA Component of Armour Fraction II of BGG estimated as 8–16% of total.

[f] In the presence of Freund's complete adjuvant (alone) or in previously immunized mice.

[g] Quantities in rabbits *adjusted* to be comparable to mice on a dose-per-body-weight basis.

[h] Estimated from preliminary data and by analogy with data for other γ-globulins.

TABLE IV
Low-Zone Paralysis: Species, Proteins, and Forms of Treatment That
Permit Induction of Paralysis in Dose Ranges <10 mg./kg. Body Weight[a]

Normal adults
  Mouse
    BGG (BγG), HGG, EGG, PGG, CaGG, RGG (Dresser, 1962b, and unpublished)
    BSA (Dietrich and Weigle, 1964; Mitchison, 1964)
  Rabbit
    LGG (Dresser and Gowland, 1964)
  Guinea pig
    BGG (Battisto and Miller, 1962)
    BGG, HSA (Dvorak et al., 1965)

Newborn
  Mouse
    BSA, lysozyme (Mitchison, unpublished)
  Rat
    Flagellin (Nossal and Ada, 1964; Nossal et al., 1965)

Postirradiation
  Mouse
    BSA, lysozyme, Diphtheria toxoid, RNase (Mitchison, 1967, and unpublished)
  Rabbit
    BSA (Linscott and Weigle, 1965a)

Maintenance
  Mouse
    BSA, lysozyme (Mitchison, 1967, and unpublished)

---

[a] In most cases paralysis can be detected at 0.1–1 mg./kg. body weight.

adjuvant, larger doses of poly-D-amino acid would presumably also immunize, were they not prevented from doing so by the induction of paralysis. Poly-D-amino acids are resistant to degradation in the body (Gill et al., 1965) and their potency as paralyzing agents is, therefore, reminiscent of that of the pneumococcal polysaccharides.

Schechter et al. (1964a) provide an example of a somewhat different nature. They find that poly-DL-alanine does not immunize rabbits, even when incorporated in Freund's adjuvant, but does induce a form of paralysis when administered in a series of injections commencing in newborn rabbits. The animals treated in this way fail to respond to immunization with poly-DL-alanylated proteins by producing antibody to the side chains, although normal controls do so. This indicates that the requirements for paralysis induction are less stringent than for immunization, although this may be true only in a quantitative sense.

2. Weakly Immunogenic Antigens

Once it has been established that the capacity to immunize can vary independently of the capacity to paralyze, antigens can be expected to occur which will immunize only at doses higher than the threshold for paralysis (Fig. 1). Antigens of this type we term "weak immunogens." Bovine serum albumin in mice belongs to this category (Mitchison, 1964, 1967). When injected together with adjuvant, BSA elicits approximately the same titers of antibody as strong immunogens, such as lysozyme or diphtheria toxoid. But when injected alone, a dosage can be found in which paralysis is induced but which is below the minimum level required for immunization ("low-zone" paralysis). As before, we employ the term "paralysis" to mean a state in which normal responsiveness to challenge with the antigen in adjuvant is reduced or abolished.

Quantitative exploration of the concentration of antigen required to induce paralysis has shown that the threshold is remarkably constant for a variety of antigens, circumstances, and animals (Table IV). The data shown in this table have been collected from different sources, and the schedules of antigen administration are not strictly comparable. Those proteins which are eliminated slowly (albumin and γ-globulins in rabbits, γ-globulins in mice) were given in the form of a single or small number of injections, whereas the more rapidly eliminated proteins (non-γ-globulins in mice) were given in a series of closely spaced injections. In order to bring the thresholds into line with one another, reference must be made to the dose per injection rather than the total overall quantity of antigen. The validity of this type of comparison rests on the assumption that paralysis is induced at a rate which is slow in comparison with the rate of elimination of the more rapidly eliminated proteins. These proteins are eliminated with half-lives of less than 1 day (Terres and Wolins, 1959; Dietrich and Weigle, 1963; Mitchison, 1964, and unpublished). Induction of paralysis (without the complications introduced by concomitant immunization—see Section II,C) requires a matter of weeks with microgram doses of BSA (Mitchison, 1967).

In comparison with this constancy of paralysis induction threshold, the minimum quantity of antigen required for immunization appears to be highly variable. In the context of paralysis induction, the most striking example of this type of variation is the difference between a weak immunogen, as exemplified by BSA, and strong immunogens such as diphtheria toxoid, ovalbumin, and lysozyme (Mitchison, 1967). For it is this part of the variation that explains the curious bimodal response to weak immunogens (Fig. 1). The "valley" of low-zone paralysis falls in the gap between the thresholds for immunization and paralysis induction; for those antigens which have lower thresholds of immunization, this valley is filled by the immunization "hill." Nevertheless, there are no grounds for believing that this end of the distribution of immunization thresholds is anomalous; equal variation can presumably be found among strong immunogens, but with less dramatic consequences.

Bainbridge and Gowland (1966a,b, and personal communication) have discovered a situation in which allogeneic cells appear to act as a weak immunogen in the sense here defined. They assay the response of CBA strain mice to ^{51}Cr-labeled (CBA $\times$ A)F_1 lymphocytes by measuring recovery of radioactivity from host lymphoid organs. Pretreatment of the host with intermediate numbers of spleen cells induces immunity, i.e., recovery of radioactivity is reduced compared with that of untreated controls (Fig. 2); pretreatment with either very small or very large numbers appears, according to the same criterion, to induce paralysis.

Paralysis can be readily induced to some haptens in adult guinea pigs. Such a state of paralysis is recognized by the inhibition of immunity in the form of a specific, delayed, skin hypersensitivity reaction which is induced by topical or intradermal application of the hapten to normal guinea pigs. Skin hypersensitivity of this kind is greatly enhanced by immunization of the animal by protein–hapten conjugates in Freund's adjuvant. Intravenous injection of neoarsphenamine (Sulzberger, 1929, 1930) or feeding of haptens such as DNCB or picryl chloride (see reviews by Chase, 1959; Chase and Battisto, 1959) leads to a state of paralysis as shown by a specific inhibition of immunity (Sulzberger-Chase phenomenon).

It seems likely that these haptens, which are active in the Sulzberger-Chase phenomenon, conjugate spontaneously to autologous proteins to form hapten–protein conjugates. In one view the conjugates formed after feeding would be nonimmunogenic and would induce paralysis, whereas hapten–dermal protein conjugates, for instance, would be immunogenic and would immunize the guinea pigs.

Frei et al. (1965) have suggested an alternative mechanism for the Sulzberger-Chase phenomenon. The suggestion is that in a heterogeneous mixture, the immunogenic hapten–protein conjugates are screened out of the circulation by liver macrophages and that only nonimmunogenic material is left to reach the lymphoid tissues, where paralysis is induced. Battisto and Miller (1962) showed that hapten–protein conjugates injected via the hepatic portal system were effectively nonimmunogenic and induced paralysis, but when injected by other routes were immunogenic and immunized. The hypothesis of Frei *et al.* implies that antigen in liver macrophages is either not immunogenic or is not available for immunization of lymphoid cells; the possible role in the immune response,

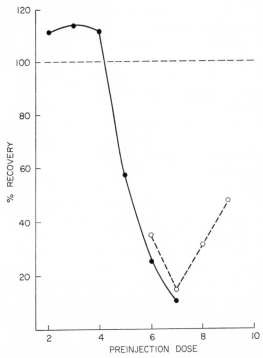

FIG. 2. Dosage zones for histocompatibility paralysis (Bainbridge and Gowland, 1966a,b, and personal communication). Paralysis measured by the recovery of labeled lymphocytes in the lymph nodes of pretreated mice. Percentage recovery of ^{51}Cr-labeled F_1 lymph node cells, 24 hours after intraperitoneal injection into CBA mice, where 100% is the recovery in control CBA mice that had not been pretreated. The preinjection dose units are $\log_{10}$ lymph node cells injected intraperitoneally ($\bullet$), and intravenously ($\bigcirc$), 1 week prior to the injection of labeled cells.

of antigen processed by perifollicular macrophages, is discussed in Section VI. Lowney (1965) showed that several topical applications of skin-sensitizing haptens (DNCB, NDMA) induce a delayed skin reaction but, at the same time, induce a state of paralysis to later challenge. This result does not help to decide between the two hypotheses outlined above but does fit in with what we have to say about concomitant immunization during the induction of paralysis, in adult animals, by immunogenic antigens (Section II,C).

Jones and Leskowitz (1965) showed that a conventional immunological paralysis could be induced to an arsanilic acid–protein conjugate by injection into neonatal guinea pigs. A similar but shorter-lived effect was seen when an arsanilic acid–tyrosine conjugate was used. This work has been extended to adult guinea pigs (Leskowitz, 1967), where 3×10^{-5} mole of the hapten–tyrosine conjugate could suppress a delayed skin reaction 1 week later. The effect may be akin to the Sulzberger-Chase phenomenon; this view is supported by the observation of Borek et al. (1965) who showed that 3-azobenzene arsonate hexa-L-tyrosine conjugates spontaneously to guinea pig proteins under physiological conditions in vitro. An alternative hypothesis is that the molecular size requirements, defined by Landsteiner (1936) for immunization by haptens, may be less stringent for paralysis induction.

An exception to the foregoing scheme must now be mentioned. Gell and Kelus (1966) find that after exposure to large amounts of maternal allotype (γ-globulin isoantigen) via normal intrauterine transfer in rabbits, neither immunity nor paralysis can be detected. A convincing explanation of this anomaly is lacking.

3. Heterogeneity of Antigen

Several studies indicate that a partially denatured or aggregated minor fraction of purified protein is responsible for immunization and that, if this fraction can be removed, the paralyzing capacity of the remaining major fraction is revealed. It is well established that animals injected with a heterogeneous mixture of proteins may become immunized by some and paralyzed by others (Dixon and Maurer, 1955; Shaul, 1962, 1963). In vivo screening, i.e., passage of material through the circulation of an animal, can be used to separate BγG from BγA in a mixture in BGG (Dresser, 1963); BγA is removed during passage as a consequence (1) of having a shorter normal half-life, and (2) of provoking a specific immune response. An indication that heterogeneity of the same type can be detected within samples of purified protein was observed by Dresser (1962b). High-speed centrifugation (20,000–30,000 g) removed the capacity of BγG to induce sporadic immunization and thus enabled the paralyzing capacity of low doses of this antigen to be discovered.

The hypothesis that the importance of screening procedures lies in the removal of material subject to phagocytosis was first advanced by Frei et al. (1965; see also Frei, 1964). They found that BSA which had been passaged *in vivo* for 2 days lost its capacity to immunize but retained a detectable capacity to paralyze. Further discussion of the role of phagocytosis is reserved for Section VI. There is some doubt about the extent to which the phenomenon of *in vivo* screening can be repeated with other albumins and in other animals. Experiments of a similar nature did not show that the capacity of HSA to immunize rabbits could be removed by passage (J. H. Humphrey and Y. Auzins, personal communication). Furthermore, passage of BSA through rabbits did not alter the ability of the protein to immunize mice (H. H. Wortis, personal communication). Attempts to demonstrate the same effect with BSA in mice have failed (Mitchison, 1964; Frei *et al.*, 1965). It seems likely, therefore, that Frei *et al.* were fortunate in their choice of material, and perhaps even in the exact extent of aggregation in their batch of protein.

C. Antigen Overloading: High-Zone Paralysis

The paralyzing effect of antigen excess was discovered by Glenny and Hopkins (1924) who injected EGG (antibody to diphtheria toxoid) into adult rabbits and followed its elimination from the circulation. They demonstrated, in two rabbits, that continued injection of the antigen resulted in a marked decrease of immune elimination. Interest in paralysis of this type was revived when Dixon and Maurer (1955) showed that adult rabbits could be rendered specifically unresponsive by prolonged treatment with a variety of foreign plasma proteins. Subsequent examples of paralysis induced by large doses of protein antigen in adult animals have been reviewed by Smith (1961; see also Dresser, 1962a, 1965; Sercarz and Coons, 1963; Mitchison, 1964; Černý *et al.*, 1965). Parallel studies with transplantation antigens have shown that paralysis can be induced during adult life by the administration of doses of cell-free extracts that are massive in comparison with the doses required either to immunize adults or to paralyze the newborn. Gowland (1965) and Good *et al.* (1966) have reviewed recent transplantation work on this subject, and Billingham *et al.* (1965) have made a detailed study of paralysis by the Y-linked transplantation antigen.

The relationship between this high-zone paralysis and the low zone is shown diagrammatically in Fig. 1, and is assessed quantitatively for γ-globulin antigens in Table III. Another assessment, for BSA in mice, has been made by Mitchison (1967), and there is agreement that there is a difference of dose requirement of the order of $\times 10^4$. An apparent exception to the rule that massive doses of antigen are required for paralysis of adults by an immunogenic antigen can be found in the case of pneumococcal polysaccharide. Part of the data summarized in Table III shows that paralysis can be induced by as little as 50 μg. of Type III pneumococcal polysaccharide in adult C3H mice (only 5 μg. required for Sw. white mice). By comparison with the dose of immunogenic protein antigen required to achieve the same relative inhibition of the immune response, these amounts seem very small. The fact that pneumococcal polysaccharide is not digested by any mammalian enzymes may account for the small amounts required for paralysis (Felton and Ottinger, 1942; Felton, 1949; Felton *et al.*, 1955a; Stark, 1955b). An alternative hypothesis is offered in Table III (see also Dresser, 1963) which explains the apparent difference by listing the quantities of antigen in terms of *determinant units* (see Table III for explanation); in these terms the quantities are very similar.

The question whether paralysis in adult life is more easily induced in respect of "weak" antigens has become involved. Clearly, nonimmunogenic and weakly immunogenic antigens, in the sense used in this review, can be used to induce paralysis in an unusually economical manner, for with them the high-zone of dosage can be avoided altogether. Among antigens that immunize at or below their paralyzing dose, a graded series of comparable antigens can sometimes be found which vary from weak antigens that paralyze normal adults fairly easily to strong antigens that can paralyze only in young animals or not at all. The best evidence of such a series comes from studies of transplantation immunity (Gowland, 1965). Purified protein antigens cannot apparently be arranged in a similar series. Lysozyme, in terms of our present classification, is a strong antigen in mice and immunizes them when administered in doses below the paralysis-inducing threshold; yet the high zone of paralysis is entered at about the same dose level as is required for the weakly immunogenic BSA and the same is probably true of diphtheria toxoid which is an immunogen potent at even lower doses than lysozyme (Mitchison, 1967, and unpublished data). Presumably the factor that is principally responsible for establishing the position of the threshold of high-zone paralysis, in respect of these protein antigens, is the level of antibody attained by an animal during its immune response; this does not appear to vary much from one protein antigen to another. Different proteins, therefore, resemble one another more closely in their capacity to induce high-zone paralysis than might be expected.

When the rates of entry into high-zone paralysis induced by immunogenic antigens are examined in detail, two characteristic features can be found. To start with, animals display a stage during which immunity can be detected before they become fully paralyzed (Glenny and Hopkins, 1924; Brent and Gowland, 1962, 1963; Siskind and Howard, 1966; Nossal and Austin, 1966). It follows, therefore, that studies such as those of Crowle (1963), Frey et al. (1964), Dorner and Uhr (1964), Dresser (1965), and Claman and Bronsky (1966), in which immunity is deliberately induced prior to paralysis, do not differ essentially from other studies of high-zone paralysis. Later on, as the animals become paralyzed, responsiveness declines in an exponential manner (Mitchison, 1962c, 1964; Dresser, 1962b, and unpublished data). In an extreme case the decline in responsiveness can be followed over a period of several weeks; kinetics of this order, it has been argued, imply the gradual recruitment or loss of cells (Mitchison, 1961). Concomitant immunization declines in parallel with responsiveness, after its initial rise. Stages in the rise and fall of this concomitant immunization can be detected by releasing mice during a course of injections with BSA and measuring the quantity of antibody that they produce spontaneously (Mitchison, 1964). Day and Farr (1966) have shown that rabbits given a continuous dose of 500 mg./day of BSA for 2 months still synthesize an antibody which can be detected by a binding test with HSA; alternatively, the serum can be fractionated to remove excess of the injected BSA, before being tested for nonabsorbed anti-BSA antibodies.

In attempting to interpret these observations, it seems clear that there is very little to suggest that the eventual state of paralysis which can be achieved under these conditions differs radically from low-zone paralysis. The stability of paralysis induced by large doses of BSA (Mitchison, 1965a), for example, is of the same order as that induced by low doses of BGG (Dresser, unpublished data). A hypothesis which postulates a susceptibility to paralysis induction applying only to immunized cells can, therefore, be accepted only with the utmost reluctance. The hypothesis of "allergic death" advanced by Jensen and Simonsen (1963) belongs to this category; cells that engage in antibody production are assumed to develop a special susceptibility to lysis upon contact with antigen.

An effect produced by injection of HSA into chickens does raise difficulties (Černý et al., 1965). The injection of a large dose causes a delay in the appearance of both circulating antibody and cells in the spleen which display a positive reaction to a specific immunofluorescent test for antibody. The delay, which is a matter of a few days, is too short to be interpreted as paralysis, and quantitative considerations suggest that the immune response is not merely being masked by excess of antigen. An experiment in which massive amounts of BGG were injected into immunized mice led to a similar conclusion (Dresser, 1962a). An adequate explanation of the phenomenon is lacking.

If it is accepted that the mechanism of induction involved in high- and low-zone paralysis is fundamentally similar, the question arises why the quantities of antigen required are so different. Two answers, both of which lack strong supporting evidence, may be offered. In the first place, concomitant immunization may simply mask paralysis, which is, nevertheless, induced covertly. As a consequence the recruitment of new cells into the immune response is blocked, and after a time the cells which were initially committed to immunity, and their progeny, die out. "Terminal differentiation" of this nature is postulated by Sterzl (1966). In the second place, the antibody produced during concomitant immunization may neutralize antigen and thus raise the quantity of antigen required for induction of paralysis. The hypothetical effect is thus analogous to the control by antibody of its own synthesis, at least if this control is interpreted as due to neutralization (Uhr, 1967). Furthermore, antibody administered passively to the newborn inhibits the induction of paralysis by a bacterial antigen (Friedman, 1965c). The objection might be raised that insufficient antibody is produced to account for the demand for antigen. In BSA high-zone paralysis in mice, only microgram quantities of antibody are released into serum during the course of paralysis, and the clearance of antigen is not detectably accelerated (Mitchison, 1964, and unpublished). The macrophage surface seems a more acceptable site for neutralization; for discussion of the evidence for macrophages from analogous immunization studies, see Uhr (1967). Alternatively, Dresser (1963, 1965) suggests that the immunologically competent cell or its immediate descendant, may be a possible site for neutralization of antibody competing for antigen on receptors.

D. Suppression of the Immune Response

In addition to the variation in immunogenic capacity between antigens which has so far been the principal subject of discussion, the responsiveness of the host can also be varied. Treatments which reduce the immune response by definition tend to render antigens nonimmunogenic, and can, therefore, be expected to reveal low-zone paralysis. Administration of antigens to the newborn is a special case of this kind of treatment (Section II,A). Other treatments which have been widely used for this purpose include irradiation, immunosuppressive drugs, and to a lesser extent prior induction of specific unresponsiveness (i.e., maintenance of paralysis).

The nonspecific suppressive action of radiation and drugs has been reviewed by Taliaferro (1957), Stoner and Hale (1962), Makinodan *et al.* (1965a), Calabresi and Welch (1965), Humphrey (1965), and Schwartz (1966). The review of Schwartz deals with studies in which treatment with immunosuppressive agents was combined with administration of an antigen, and with the effect of this treatment on a subsequent test of reactivity to the antigen. Ample documentation concerning strongly immunogenic antigens was offered for the proposition that immunosuppressive treatment greatly reduces the dose required to paralyze adult animals. In this respect, no clear distinction can be drawn between metabolic and mitotic inhibitors. Detailed studies of the threshold dosage requirements for paralysis after immunosuppression are referred to in Table II. They have shown quantitatively that the same threshold obtains as in the newborn, and as in normal adults for weak and nonimmunogenic antigens.

Other forms of immunosuppressive treatment which might be expected to facilitate the induction of paralysis in adults have not been fully investigated from this point of view. They include: (*a*) chronic drainage from the thoracic duct (McGregor and Gowans, 1963, 1964), (*b*) administration of serum α_2-globulin (Kamrin, 1959, 1966; Mowbray, 1963a,b; Mowbray and Hargrave, 1966; Mowbray and Scholand, 1966), and (*c*) administration of antilymphocyte serum (Woodruff and Anderson, 1963; Levey and Medawar, 1966). In the case of antilymphocytic serum the possibility must be borne in mind that induction of paralysis may be prevented, as well as immunization.

Prior induction of specific unresponsiveness can be regarded as another means of preventing immunization and thus facilitating induction of paralysis. It might seem paradoxical to measure the power of an antigen to paralyze in an animal that is already specifically paralyzed, but this is not necessarily so. Recovery from paralysis normally proceeds spontaneously (Section V), and so it is possible to measure the dosage of antigen required to maintain a pre-existing state of paralysis. Two independent studies have shown that smaller doses of strongly immunogenic antigen are required to maintain than to induce paralysis; one study was with mouse erythrocytes in the rat (Mäkelä and Nossal, 1962), and the other was with allogeneic erythrocytes in the chicken (Mitchison, 1962c). On the other hand, the dose of BSA or lysozyme required to maintain paralysis in the mouse is not significantly different from that required to induce paralysis in the low zone (Table IV).

III.   Related Phenomena

A.   Maternal Paralysis by Fetal Antigens

In certain circumstances fetal isoantigens paralyze mothers. For instance, BALB/c mice become able to support the growth of a DBA/2 tumor, which they would normally reject, after pregnancies resulting from matings to DBA/2 males (Breyere and Barrett, 1960). Other examples have been found in mice (reviewed by Breyere, 1964), and a similar phenomenon has been described in the armadillo by Anderson and Benirschke (1964). The mechanism of paralysis by fetal antigens is not known; presumably there must be contact between lymphoid cells of the mother and fetal antigen. In the normal way grafts bearing the antigens in question provoke an immune response, and so the maternal fetal relationship can be regarded as one which renders potentially immunogenic antigens nonimmunogenic.

B.   Competition of Antigens

An immune response to an antigen may be reduced if another antigen is injected at approximately the same time. This phenomenon of antigen competition has been comprehensively reviewed by Adler (1959, 1964). The inhibitory effect is nonspecific and is not dissimilar to the effect seen after the administration of drugs or irradiation, and might on similar grounds be expected to facilitate paralysis (see Section II,D). There is no direct evidence to support this view, although the experiments of Liacopoulos and co-workers are compatible with it. Liacopoulos (1961), Liacopoulos et al. (1962), Liacopoulos and Neveu (1964), and Liacopoulos and Perramant (1966) have shown that massive doses of protein antigens administered to guinea pigs, rats, mice, and rabbits result in short-lived inhibition of the immune response to unrelated antigens. A likely explanation of the phenomenon is that the massive dose of antigen provokes an immune response, with concomitant paralysis (Section II,C), and that it is this immunization which competitively inhibits the response to the second antigen. The paralytic effect of the massive dose, which is a prominent feature of the experiment, need not necessarily be relevant to the nonspecific immunosuppressive action.

C. HETEROGENEITY OF THE IMMUNE RESPONSE

The immune response is heterogeneous, involving both cellular (delayed hypersensitivity, homograft reaction) and humoral elements (circulating antibodies). Furthermore, at least four distinct classes of circulating antibody can be distinguished in guinea pigs (γM, γA, γG$_1$, γG$_2$) (White et al., 1963; Benacerraf et al., 1963), and five in mice (γG$_2$ subdivided into A and B) (Fahey et al., 1964). In addition to having distinct antigenic and chemical characteristics the different classes can also be distinguished in many instances by function; for example, γM antibodies readily fix complement and are highly cytolytic, whereas some other classes fix complement less readily and are weakly or not at all cytolytic. Consequently the detection of an immune response depends on both (a) the nature of the assay procedure, and (b) the type of response under study.

A state of paralysis can, therefore, be detected on the basis of one assay, where another would indicate immunity. The phenomenon of "immune deviation," described by Asherson and Stone (1965) and Asherson (1966) provides an example of this. These authors show that preimmunization of guinea pigs by an alum-precipitated protein antigen causes the animals to give only a γG$_1$ response to challenge with the same antigen in complete Freund's adjuvant. Control animals which had not been preimmunized show both γG$_1$ and γG$_2$ antibodies and also manifest delayed hypersensitivity. Their experiments are an extension of a previous discovery that immunization of guinea pigs and mice with a protein antigen in complete Freund's adjuvant elicits a γG$_1$, γG$_2$, and delayed hypersensitivity response, whereas immunization with incomplete adjuvant elicits only a γG$_1$ response (White et al., 1963; Benacerraf et al., 1963; Coe and Salvin, 1964; Askonas et al., 1965; Barth et al., 1965; Coe, 1966; Wilkinson and White, 1966).

Something akin to "immune deviation" may underlie paralysis of the type described by Battisto and Bloom (1966a,b). They showed that injection of the picryl group or BGG coupled onto sheep red blood cell stromata rendered guinea pigs unresponsive to challenge with the antigens in adjuvant. The methods used to detect the absence of an immune response were systemic anaphylaxis, passive cutaneous anaphylaxis, and delayed skin hypersensitivity. In contrast, Chou et al. (1966) found that HSA coupled onto nucleated syngeneic thymocytes proved highly immunogeneic in rabbits, whereas similar quantities of HSA in solution induced neither immunity nor paralysis. Chou et al. (1966) used passive hemagglutination and precipitation in agar gel to detect immunity. Borel *et al.* (1966) using a hapten, and Crowle and Hu (1966) using a protein, have recently reported phenomena which could either be interpreted as immune deviation or, alternatively, as situations where only one part of the immune response is paralyzed.

The mechanism of immune deviation is not understood. One possibility involves "feedback" inhibition of antibody synthesis by antibody (Kaliss, 1958; Uhr and Baumann, 1961) or, more specifically, by one antibody class of the induction of the synthesis of another (Sahiar and Schwartz, 1964, 1965, 1966; Finkelstein and Uhr, 1964). Asherson's (1966) experiments indicate that this mechanism is unlikely, at least as far as inhibition of a delayed hypersensitivity reaction by passive antibody is concerned.

D.   SPLIT TOLERANCE

Brent and Courtenay (1962) injected (CBA × C57BL/6)F_1 cells into neonatal A strain mice and were later able to show that the A strain mice were paralyzed by CBA antigens but not by all the C57BL/6 antigens. Experiments of the same sort were carried out by Krenova-Peclova *et al.* (1963), who obtained similar results. Billingham *et al.* (1965) and Weissman (1966) have carried out detailed investigations of the phenomenon using the male (Y-linked) histocompatibility antigens in inbred mice. "Split tolerance" may be the result of heterogeneity of response to a mixture of antigens of heterogeneous strength; i.e., in Brent and Courtenay's experiments the C57BL/6 histocompatibility antigens of the F_1 cells are more immunogenic in neonatal A strain mice than are the equivalent CBA antigens.

IV.   Specificity of Paralysis

Studies of immunological paralysis normally include controls in which nonspecific damage to the mechanism of response is assessed by immunization with one or more unrelated antigens. These should provoke an unimpaired response, provided that the antigen used does not cross-react with the original paralysis inducer. The state of paralysis should at the same time be left unaffected. Thus, for example, mice paralyzed with Type I pneumococcus polysaccharide and then immunized with Type II or III become resistant to challenge with living organisms of Types II or III, but not to challenge with Type I (Felton *et al.*, 1955a). Similarly, mice tolerant of skin from one inbred strain reject homografts from other strains (Billingham *et al.*, 1956), and rabbits tolerant of one foreign protein react normally to others (Dixon and Maurer, 1955). Due to cross-reactions between serum proteins from mammalian species, the situation here is more complex than was originally thought, as will be discussed. But later studies have amply verified the specificity of the paralysis that can be induced by purified proteins such as BSA in respect of noncross-reactive antigens (Mitchison, 1964). Indeed, the induction of paralysis by a nominally pure preparation of protein frequently causes immunization to minor components of the preparation (Dixon and Maurer, 1955).

The kind of observation just mentioned implies that the mechanism of paralysis is no less sensitive to antigenic difference than is the immune response. However, a question which must still be answered is whether the recognition unit concerned in paralysis binds antigen in precisely the same way as does antibody. We shall have, therefore, to consider what happens when paralyzed animals are immunized with antigen that cross-reacts with the original inducer, for it is this kind of experiment that suggests that the recognition unit does not behave quite like antibody. Before doing so we ought to realize the issue at stake. One of the simplest hypotheses of recognition postulates that lymphocytes bear receptors for antigen that are antibodies and that these antibodies are an accurate sample, in respect of their combining sites, of the kind of antibody that the cell or its progeny can make during an immune response (whether one cell has one or more kinds of receptor is not strictly relevant here). Combination of a receptor with its antigen signals the cell to commence multiplication and synthesis of antibody, provided that the antigen is presented to the receptor in a suitable (but for the present unspecified) manner; presentation in another manner leads to irreversible blockade of the site or even, according to one attractively simple view, to cell death. The two modes of presentation correspond respectively to immunization and to paralysis. It is this hypothesis of the *accurate sample* that is at stake in the present discussion.

Naturally, paralysis is not the only area of immunology to provide relevant information. Similar queries about the recognition mechanism arise in the contexts of delayed hypersensitivity and of the secondary response, i.e., in other situations where lymphocytes interact with antigens. Very much the same suggestion has been made in all three contexts that recognition involves a wider area of antigen surface than does binding by antibody (reviewed by Mitchison, 1966).

The experiments that are relevant have been performed in three ways: (1) paralysis is induced by one protein, followed by immunization either with cross-reactive proteins or with the original protein conjugated to hapten; (2) paralysis is induced by a hapten conjugated to one protein, followed by immunization with the same hapten conjugated to another protein; (3) a normal animal, presumably tolerant of its own antigens, is immunized either with autologous denatured or conjugated antigens, or with cross-reactive tissue from another species. Work of the third type, concerned with autoimmunity, tends to gather in a separate literature. The antigens involved in autoimmunity are relatively ill-defined, and the relationship between natural and acquired tolerance is still in doubt. One of the authors has already reviewed this subject (Mitchison, 1966), and we shall not refer to it further here.

Cinader and Dubert (1955, 1956) paralyzed rabbits with HSA and subsequently immunized them with sulfanyl–azo-HSA, and noted for the first time that antibodies were formed not only against the new determinant but also against the original protein. Similarly, Curtain (1959) induced paralysis in rabbits with Bence-Jones proteins (i.e., human myeloma L chains) and subsequently immunized them with the corresponding intact myeloma proteins (H + L chains); again, antibody was formed, in some animals, to the L-chain determinants. Denhardt and Owen (1960) immunized rabbits paralyzed by BSA with sulfanyl–azo-BSA but failed to find antibody to BSA. The first systematic study of termination by cross-immunization was carried out by Weigle (1961, 1962, 1964a,b, 1965b) and Linscott and Weigle (1965b). The choice of a suitable degree of cross-reaction turns out to be important. No effect is produced unless the second antigen cross-reacts with the original inducer, but if the similarity is too great even the determinants that were absent on the original inducer may not elicit a response. For example, sheep serum albumin (SSA) will not terminate BSA paralysis in rabbits (normal cross-reaction of SSA antiserum with BSA is 75%), whereas HSA does so regularly (normal cross-reaction 15%) (Weigle, 1961). Among the various conjugates tested, mixed sulfanyl and arsenyl substituents proved most effective, suggesting that strong varied groups are required in the cross-immunizing antigen (Weigle, 1962). In a combination of antigens that otherwise shows only a weak effect, the use of Freund's adjuvant may help, but this is not always essential (Weigle, 1964c). The effect can be prevented by injecting the original inducer together with the cross-reactive antigen (Porter, 1962; Humphrey, 1964b; Weigle, 1964c), an observation that at present lacks satisfactory explanation.

Variations on the foregoing design of experiment have yielded similar results. Nachtigal and Feldman (1964) induced paralysis by a combination of irradiation and prolonged administration of antigen in adult rabbits and, again, found that immunization with hapten–protein conjugates can terminate paralysis. Schechter et al. (1964b) examined the consequences of cross-immunization with peptidyl proteins and found that strongly immunogenic groups (tyrosyl) were more effective in terminating than weak ones (alanyl); they again found, by varying the degree of enrichment, that an intermediate degree of cross-reactivity was the most effective.

Despite some indication to the contrary, proteins for which paralysis has been induced and autologous proteins appear to act in much the same way as hapten-carriers. Boyden and Sorkin (1962) recorded a difference; sulfanyl–azo-RSA raised antihapten antibodies in normal rabbits, whereas sulfanyl–azo-HSA did not do so in rabbits paralyzed with HSA; but this has not been confirmed in subsequent experiments (Nachtigal and Feldman, 1964). Weigle (1965a) was able to elicit antibodies to native thyroglobulin by immunization with conjugated thyroglobulin.

When paralysis induced by one protein is terminated by immunization with a conjugate of the same protein to a hapten, the two antigens have in common at least some of their protein determinants. The situation can be inverted by paralyzing with one conjugate and then immunizing with a second conjugate of the same hapten with another carrier; in that case the antigens have in common the haptenic group. Experiments designed in this way at first gave no indication of any appreciable paralysis directed toward the haptenic group. Coe and Salvin (1963) found that guinea pigs that had been paralyzed by gastric feeding of DNCB responded normally to most other DNP conjugates. 2,4-Dinitrophenyl conjugated to autologous serum protein did not provoke a normal response in the paralyzed animal, but this could have been due to the presence of similar conjugates in the original inducing antigen. The doubt was resolved when Weigle (1965b) found that rabbits paralyzed by picryl–BSA did not respond normally to the picryl group presented in the form of a conjugate with BSA, although they did so with other carriers. Further evidence of paralysis directed toward a single determinant on nonautologous carriers has been obtained with NIP (Brownstone et al., 1966). The anti-NIP response induced by NIP–CG is delayed in mice paralyzed with NIP–BSA; but here again the effect is more marked when NIP is presented on an autologous carrier (Mitchison, unpublished). The failure to terminate with autologous carriers falls in line with what has already been referred to with nonconjugated proteins: termination occurs only when a strongly immunizing antigen and immunizing procedure is used. These experiments with single haptens provide the first direct clue to the size of the unit of antigen initially recognized; they suggest that recognition is directed not toward the whole antigen but toward some part of it not much larger than the individual hapten.

Once it has been shown that antibodies can be evoked that react with the original paralysis-inducer, several questions arise about the details of their specificity. Above all, we need to know whether they are directed toward structures in the antigen which could reasonably have been expected to react with the recognition mechanism at the time of induction. If this is not the case, the effect loses much of its interest. Weigle (1964b) and Linscott and Weigle (1965b) argue that potentially immunogenic structures present in normal protein antigens have antibodies made against them as a consequence of cross-immunization with conjugates in paralyzed animals. The evidence they present is (1) rabbits paralyzed with BSA and then cross-immunized with related proteins give an immune response to subsequent injections of BSA, although the risk of reparalysis by this treatment is considerable, and (2) the antibodies produced under these circumstances, as judged by affinity and inhibition criteria, tend eventually to resemble, more or less closely, normal anti-BSA antibodies. Nevertheless, there are difficulties. Antibodies that are directed solely toward the original protein determinants emerge only late in the course of cross-immunization, if at all. Before they appear, the animal makes antibodies with carrier-specificity toward the original protein which require the presence of the new determinants before they will precipitate (Nachtigal et al., 1965) or bind (Linscott and Weigle, 1965b). In other words, these antibodies are directed toward a mixed or "link" site; possibly, as Singer (1964) suggests, this is the hapten plus some neighboring part of the protein that it has bent over and adsorbed onto.

The implication so far is that the recognition mechanism is directed toward antigenic molecules as a whole rather than toward the kind of individual determinants that react with antibody. Initial recognition might, therefore, be brought about by a mechanism that does not involve antibody in its operation; an idea of this sort was put forward by Weigle (1961). After an antigen has been accepted at this stage, it would then be searched for individual determinants against which antibodies would be made (the search might be carried out by receptor-type antibodies). Paralysis would affect only the first step, so that determinants which got past it by virtue of their presence in an otherwise "foreign" molecule would evoke antibody, even though they were also present on the molecule used to establish paralysis. This view, in its simplest form at least, implies abandoning the *accurate sample* hypothesis. It is not difficult, however, to imagine an *ad hoc* scheme which would accommodate both the data on termination and this hypothesis. For example, the receptors of the paralyzed animals might recognize initially only those determinants that are new; once antibody production has commenced the situation alters; an antigen would be concentrated by macrophages; normally ineffective determinants could then reach the local concentration required to immunize. Normally ineffective determinants of this type could not be expected to have induced paralysis nor to immunize unless associated with determinants capable of being immediately recognized.

An example has recently been found of two antigens which display a relationship in paralysis experiments which is different from anything that has so far been described. Rats paralyzed by injection of *Salmonella* flagellin *fg* become unable to react against two other flagellar antigens, *i* and *d*, that are made in the same bacterium, but are not cross-reactive with *fg* by any of the serological tests that have been applied (Austin and Nossal, 1966). Immunization with *fg* flagella depresses the subsequent response to *i* flagella, in a manner reminiscent of Asherson's phenomenon of immune deviation (1966; see Section III). These observations cannot easily be reconciled with the accurate sample hypothesis.

V. Recovery from Paralysis

At the time of its discovery paralysis appeared to be a stable condition from which animals did not recover spontaneously, or did so only over periods commensurate with the life span. In retrospect the systems then under study can be seen to have favored stability. Either viable cells were included in the inducing inoculum, which therefore provided a continuing source of antigenic stimulation (Billingham *et al.*, 1956), or polysaccharide antigens were employed which are resistant to intracellular degradation (Felton *et al.*, 1955a). As attention shifted to proteins and nonreplicating cells, it was realized that immunological responsiveness returns some time after administration of antigen is discontinued. The speed and regularity with which this return can be observed was emphasized in the previous review of the subject in this series (Smith, 1961). Since then it has become clear that the return is not always either rapid or complete, even with protein antigens. For example, rabbits exposed to foreign serum albumin antigens at birth may remain paralyzed without further administration of antigen for over 2 years; even over such a prolonged period, however, a trend toward recovery has been detected (Bussard, 1962; Humphrey, 1964a).

In attempting to understand the mechanism of recovery, the importance of several factors has to be assessed. These include (1) catabolism of antigens: is this the rate-limiting step in recovery and, if so, is extracellular catabolism—the only kind that can easily be measured— of importance? (2) Cellular turnover: is recovery brought about by the recruitment of new cells from precursors that are not susceptible to the action of antigen and, if so, what cell types are involved? (3) Responsiveness postrecovery: is the animal that has recovered normally responsive, do special restrictions continue to apply, and does antibody production commence spontaneously? So far as the mechanism of paralysis itself is concerned, perhaps the most important question that can be asked is whether any evidence can be found of an individual cell changing in state from paralysis to responsiveness. Failure to detect such a change would be strong evidence against the existence of paralyzed cells and, hence, evidence in support of the view that paralysis involves the death of cells precommitted to the synthesis of antibody of a particular specificity.

RATE OF RECOVERY

The process of recovery of an animal from a state of immunological paralysis, and the difficulties that beset any attempt to define a rate are illustrated in the example shown in Fig. 3. Recovery evidently proceeds slowly and requires a large number of animals for systematic measurement. For this reason the most complete studies have been carried out on mice. A study of this type is basically an attempt to assess the rate at which the number of immunologically reactive cells increases (or possibly the increase in reactivity of individual cells), but the serological measurement of response to a challenge immunization does this only indirectly. The quantity of antibody evolved is probably directly proportional to the number of responsive cells over at least the lower part of the range early in the response; this relationship has been found to hold good for responsive cells transferred in small numbers into irradiated hosts (Makinodan and Gengozian, 1960; Mäkelä and Mitchison, 1965a). Short of a direct measure of the number of responsive cells, the proper scale to use is fractional recovery of the normal response. This is often not computed, or else measured on a scale that may not be directly proportional to quantity of antibody; for example, by timing accelerated clearance of labeled antigen from the circulation. One other difficulty besets comparison between different studies. The zero point from which recovery starts is not necessarily a state of complete paralysis. In the example shown, the zero is more convincing than usual because it is possible to detect a latent period before any detectable recovery occurs, yet even here absolute zero is necessarily a matter of extrapolation. Because of the uncertainty that the animals were properly paralyzed before any recovery was detected, claims of unusual recovery rates have to be treated with caution.

For these reasons the collection of data on recovery from paralysis induced by protein antigens (Table V) will not bear close quantitative comparisons. Nevertheless certain trends are clear. Recovery is detected in every case except in those early studies where the period of observation was curtailed. In the mouse, recovery proceeds more or less to

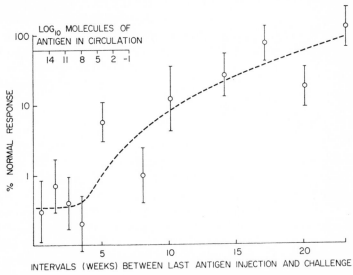

FIG. 3. Recovery from paralysis induced in adult mice by the injection of 5 mg. BSA per week for 10 successive weeks; responsiveness assayed serologically (Farr technique) with challenge immunization administered in adjuvant (Mitchison, 1965a).

completion within a few months, but more time than this is required by rabbits; there is a definite hint here that in larger and more long-lived animals, recovery proceeds more slowly. Recovery proceeds most rapidly after an initial latent period, and a modicum of unresponsiveness remains over long periods.

We turn now to a more detailed examination of the factors affecting speed of recovery.

1. Nature and Dosage of Antigen

The pneumococcal polysaccharides provide a series of antigens for which recovery appears definitely to be delayed. Recovery in both adult

TABLE V
SPONTANEOUS RECOVERY INDUCED BY PROTEIN ANTIGENS

Animals[a]	Antigen	Inducing dose (mg.)	Period of recovery (weeks)	Method of assay[c]	Source
Rabbits NB	BSA	10	15–27 ⎫	C	Smith and Bridges (1958)
		0.1	7–9 ⎭		
Rabbits NB	BSA,HSA,HGG	0.1–100	9–>112	C	Humphrey (1964a)
Rabbits AX	BSA	300	8–>16	S	Linscott and Weigle (1965a)
Guinea pigs NB	BSA	22	>14	C	Humphrey and Turk (1961)
Guinea pigs A	HSA,BGG	5	3–12	S,D	Dvorak and Flax (1966)
Mice NB	BSA	50	8–17	S,A	Terres and Hughes (1959)
Mice NB	BGG	1	2–6 ⎫	S	Thorbecke et al. (1961)
	BGG	10–30	7–15 ⎭		
Mice NB	R,H,S,B,E,TGG	20	4–>18 ⎫	S,C	Dietrich and Weigle (1963); Dietrich (1966)
	R,H,C,S,B,ESA	20	4–18 ⎭		
Mice A	BGG	1.2[b]	3–19	C	Claman and Talmage (1963)
Mice A	BSA	5[b]	3–24 ⎫	S,C	Taylor (1964)
	BGG	0.25	4–24 ⎭		
Mice A	BSA	5[b]	4–24	S	Mitchison (1965a)
Chickens NB	TGG	10	6–12	C	Tempelis (1965); Phillips (1966)

[a] NB, newborn; A, adult; X, irradiated.

[b] Repeated doses.

[c] A, anaphylaxis; C, clearance rate; D, dermal reactivity; S, serology.

(Felton *et al.*, 1955b) and newborn mice (Siskind *et al.*, 1963) is particularly well documented. The nature of the assay (survival of an otherwise lethal challenge) does not permit detailed comparison with data obtained with serum protein antigens but it is clear that an altogether longer time scale is involved. For example, following induction of paralysis in adult mice, recovery can be detected after 18 months, but only for Type I polysaccharide and not for Types II and III and even, then, only to a limited degree. There is general agreement that this special property of the pneumococcal polysaccharides can be ascribed to their long retention in tissues (Kaplan *et al.*, 1950; Felton *et al.*, 1955b; Stark, 1955a,b), although it is no longer believed that antigen retained in this way acts as a sink for antibody that is synthesized (Brooke and Karnovsky, 1961; Neeper and Seastone, 1963).

Unusually fast or slow recovery has not yet been convincingly demonstrated for other types of antigen, although it might be supposed that the process would occur more rapidly with antigens bearing a large number of determinant groups. One study (Dietrich and Weigle, 1963) describes considerable differences between heterologous proteins in the rates of recovery, but it is not certain that the same degree of paralysis and concomitant immunization was achieved in every case. This is the same objection that mars other studies in which recovery has been claimed to proceed more rapidly after induction with small doses (Thorbecke *et al.*, 1961; Terres and Hughes, 1959). One is left with the impression that simplifying the structure of an antigen, or increasing the inducing dose, can do little to increase the stability of paralysis, once an adequate means of induction has been found; at least the effect that can be produced is trivial in comparison with that exerted by altering the rate of cellular recruitment, as will be described.

2. Persistence of Antigen

At a time when an individual would otherwise recover reactivity, a state of paralysis can be maintained by further injection of antigen; in this sense, maintenance of paralysis depends on persistence of antigenic stimulation (Smith and Bridges, 1958; Mitchison, 1959; Terres and Hughes, 1959). It could at first be argued that recovery took place as soon as free antigen disappeared from the circulation (Smith and Bridges, 1958; Smith, 1961), and calculations were made of the number of molecules of antigen present in the circulation at the time of recovery. This view has now been abandoned, principally for two reasons. (*1*) In some circumstances an interval occurs between loss of free antigen and recovery of reactivity. The interval has to be quite long in order to make the necessary extrapolations convincing. Erythrocytes, which suffer linear rather than exponential clearance from the circulation, provide a particularly suitable type of antigen for this kind of investigation; with allogeneic erythrocytes, gaps of several months between clearance and recovery have been obtained (Mitchison, 1962b). Paralysis that endures long after clearance of protein antigens has also been found (Humphrey, 1964a; Mitchison, 1965a). (2) Marked variation in the rate of catabolism of proteins can be found that is not reflected in the rate of recovery from the paralysis that they induce. This is shown particularly clearly by a comparison between slowly cleared γ-globulin and rapidly cleared serum albumin antigens in the mouse (Dietrich and Weigle, 1963).

Assessment of the presence of antigen within cells is a much less certain matter. In order to obtain high specific radioactivity, resort has been had to external labeling with ^{125}I, and the possibility of de-iodination must, therefore, be borne in mind. Mitchell and Nossal (1966) found that BSA and flagellin had disappeared entirely from the lymphoid organs long before rats that had been paralyzed by injection of these antigens at birth recovered responsiveness. They employed bulk scintillation counting and radioautography for detection, and could detect 1–4 molecules of antigen per cell. In a less systematic study of paralysis induced in adults, in which bulk scintillation counting was used, no BSA could be detected in the lymphoid organs nor any leukocytes of paralyzed mice (Mitchison, 1965b); the limit here was 10 molecules average per cell.

3. Age and Thymectomy

Recovery from paralysis proceeds more rapidly in young than in old animals. The effect of age was discovered in chickens recovering reactivity to allogeneic erythrocytes, after paralysis induced initially by transfusion at hatching and then maintained by repeated transfusions (Mitchison, 1962b). The speed of recovery declines progressively with increase of age, with a final difference of over tenfold between the oldest and youngest birds tested (Fig. 4). Age influences the recovery of mice from paralysis by protein antigens in the same way (Mitchison, 1965a). For these antigens a single induction procedure was used, irrespective of age, thus answering any doubts that the age effect detected before might have been due to accumulation of larger amounts of antigen in the older animals that had received treatment over longer periods.

Nothing that is known of antigen catabolism fits such a pronounced age effect. The rate of elimination of erythrocytes is not appreciably faster in young animals (Mitchison, 1962a). The effect of age upon the rate of elimination of foreign proteins from the circulation has been carefully studied (Deichmiller and Dixon, 1960; Humphrey, 1961; Pace and Dresser, 1961) and a decline in rate during maturation noted; this decline, however, is much less than the decline in recovery rate. So far as intracellular catabolism is concerned, a straightforward comparison cannot be made because, in those newborn animals that have been examined, antigen is localized in a different manner from adults (Mitchell and Nossal, 1966). The known differences are not such as would account for the decline in recovery rate.

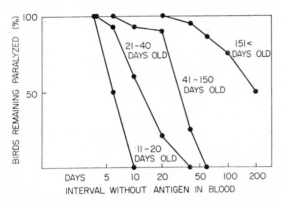

FIG. 4. The effect of age on the length of interval needed to recover from paralysis (Mitchison, 1962b). Paralysis in chickens induced by allogeneic erythrocytes; response detected by clearance of ^{51}Cr-labeled erythrocytes.

Where a parallel with the decline in recovery rate can be found is in the proliferative activity of lymphoid tissue in general and, especially, of the thymus. In the rat, for example, lymphocyte production in the thymus attains a maximum in young animals and then falls continuously over a period of 2 years, reaching a value one order of magnitude less than the peak (Andreasen and Ottesen, 1945; for review and recent references, see Miller, 1964).

The crucial test, measurement of recovery after thymectomy, has been performed by Claman and Talmage (1963) and Taylor (1964). In both series of experiments the recovery of responsiveness to a foreign serum protein antigen was examined over a period of 3 to 4 months, and similar results were obtained: recovery was markedly inhibited in thymectomized mice. Some waning of paralysis was noted even in the thymectomized animals; this seems a real phenomenon, although incomplete thymectomy, or thymic function during the brief interval between the last injection of antigen and thymectomy, have not been excluded as explanations. Little would be gained in these experiments by prolonging the period of examination post-thymectomy, for the block in recovery is already known to last nearly as long as the period of normal responsiveness after thymectomy in adult life; subsequently immunological responsiveness in general declines (Metcalf, 1965; Miller, 1965; Taylor, 1965), presumably because the existing population of reactive cells is not renewed. The block of recovery imposed by thymectomy is, therefore, virtually complete.

These experiments suggest that individual cells do not recover immunological competence during recovery of the whole animal and, therefore, that induction of paralysis involves destruction or irreversible inactivation of cells. They indicate that recovery takes place under the influence of the thymus, either involving the dissemination of reactive cells from the thymus (Ford and Micklem, 1963; Taylor, 1963) or via a thymic hormone (Osoba and Miller, 1963; Levey et al., 1963). In either case the recruitment of a new population of reactive cells is presumably involved, although the possibility that the thymus hormone enables an inactivated cell to recover has not been excluded.

4. Passive Immunization; Degradative Enzymes

The action of antiserum on paralysis attracted attention at a time when it was believed that the presence of extracellular antigen might be required for maintenance of paralysis. It was found that passive antibody had no effect on paralysis induced by BSA in chickens (Tempelis et al., 1958) or rabbits (Weigle, 1964a) or on paralysis induced by foreign erythrocytes in rats (Nossal and Larkin, 1959). An appreciable acceleration of recovery has been demonstrated upon paralysis toward allogeneic erythrocytes in chickens (Mitchison, 1962b); the effect observed could be interpreted as an advancement of the time at which the foreign erythrocytes were cleared from the circulation. The same effect can be obtained in mice with BGG, a protein that is cleared slowly from the circulation (D. W. Dresser, unpublished). An effect of this nature would not have been detected in the negative experiments that have been referred to. It can be inferred, therefore, that antigen continues to feed into the paralysis mechanism as long as antigen remains in the circulation; indeed, it is hard to imagine a mechanism of which this would not be expected. There is no hint that passive antibody can act in a more dramatic way—for example, by wresting antigen from a cellular repressive site.

An approach along similar lines was made by Brooke (1964), who injected a depolymerase specific for Type III pneumococcal polysaccharide into mice that had been paralyzed with this antigen. This procedure caused the mice to produce antibodies; as with passive antibody, however, a dramatic effect could be produced for only a few days after injection of the paralyzing antigen. We need not conclude, therefore, that the treatment affects intracellular antigen.

5. Irradiation

The cell damage and subsequent proliferation that follows irradiation might well be expected to accelerate recovery from paralysis, either through dilution-out of intracellular depots of antigen or through an alteration in the rate of recruitment of new cells. Accelerated recovery from paralysis has been found for heterologous erythrocytes in rats (Nossal and Larkin, 1959; Mäkelä and Nossal, 1962; Stone and Owen, 1963), BSA in rats (Mäkelä and Nossal, 1962), and homografts in mice (Fefer and Nossal, 1962). In these cases the paralysis was not complete; for example, foreign erythrocytes suffered fairly rapid clearance in the paralyzed animals. In other cases, where a deeper paralysis seems to have been obtained, no effect could be obtained by irradiation (Denhardt and Owen, 1960; Weigle, 1964a; Mitchison, 1965a), or one of marginal significance (Claman and McDonald, 1964). For this reason it seems likely (Weigle, 1964a) that radiation acts on partially paralyzed animals by enhancing the proliferation of responding cells, in the manner that has been demonstrated for normal animals receiving antigen prior to irradiation (Taliaferro, 1957; Dixon and McConahey, 1963). The failure to influence the rate of recruitment of new cells in this context shows how stable this process must be, for no response can be detected to the demands placed by the peripheral hypoplasia that follows irradiation.

There is another sense in which irradiation exerts an appreciable influence. Although irradiated animals that have been fully paralyzed respond to immunization no better than nonirradiated ones, their controls that have been irradiated respond poorly. Thus, by irradiating and so eradicating the existing paralyzed population of lymphoid cells, paralyzed animals can be made to approximate more or less closely to their nonparalyzed but irradiated controls (Mitchison, 1965a).

6. Postrecovery Effects

a. *Overshoot.* Immunization, in the sense of detectable production of antibody or heightened sensitivity to subsequent antigen stimulation, develops spontaneously after recovery of mice from paralysis near birth by some protein antigens (Terres and Hughes, 1959; Thorbecke *et al.,* 1961; Coons, 1963; Mitchison, 1965a) and polysaccharides (Siskind *et al.,* 1963). This overshoot of recovery was at first interpreted as a consequence of overdosage: as the quantity of antigen in control depots falls from an initially high level, a point is reached where it becomes stimulatory rather than inhibitory. Overshoot is apparently confined to circumstances where recovery takes place rapidly, for it has not been detected in older animals (Mitchison, 1962b, 1965a) or in slowly recovering species (Humphrey, 1964a).

In the light of the respective roles now assigned to macrophages and lymphocytes (see Section VI of this review), overshoot can be explained as follows. Antigen retained in macrophages serves to immunize potentially reactive lymphocytes. Following clearance of antigen from extracellular spaces at the end of a paralysis-inducing treatment, macrophages enter the period of recovery with a certain reserve of antigen which is then gradually lost. A race now starts between this process of catabolism and the recruitment of new lymphocytes under thymic influence. If recruitment is rapid, the number of new lymphocytes which undergo sensitization by the macrophage depot is large enough for immunity to become detectable. If not, the depot expires before an appreciable number of the new lymphocytes have been exposed.

An apparent paradox enters the explanation at this point: how can we account for the exceptionally prolonged character of paralysis induced by pneumococcal polysaccharide? Studies of fluorescence labeling indicate that the antigen is retained in macrophages (Kaplan *et al.,* 1950); if we assume that the antigen which can be detected in this way is responsible for maintaining paralysis, we must conclude that antigen in or on macrophages can induce paralysis as well as immunity. The question is further discussed in Section VI.

One might enquire whether enough antigen can be detected in macrophages for a mechanism of the type postulated to operate. Enough data are available for rough calculations to be made (Mitchison, 1965a). In biological terms, immunogenic relics can be detected by means of cell transfer for as long as 10 weeks after the injection of 100 mg. BSA into mice. Overshoot occurs at about the same time after termination of a course of paralysis with this antigen. In terms of radioactivity, enough

[131]I activity can be recovered from the liver and spleen at this time after injection of labeled BSA for "spontaneous" antibody production to be expected.

b. *Partial Recovery.* That recovery of reactivity is, to begin with at least, only partial has been well documented (Hašek et al., 1961; Claman and Talmage, 1963; Mitchison, 1965a). An explanation for this has naturally been sought simply in the number of reactive cells that are available for stimulation early on. Restrictions in reactivity as well as in number must now be taken into account, after the detailed study of antibodies made in previously paralyzed rabbits by Humphrey (1964b), who showed that the antisera examined (to HSA, BSA, and HGG) failed to form complexes with antigen of normal size and, therefore, failed to precipitate; furthermore, antisera to HSA, which bound appreciable amounts of intact HSA, failed to bind a 12,000-mol. wt. fragment. The observations are compatible with the presence in the antisera of antibodies directed toward a single determinant, or at least of severely restricted reactivity. In terms of the recruitment hypothesis, this suggests that the new lymphocytes are endowed with limited reactivities and are present initially in numbers small enough for errors of sampling to occur. This view has far-reaching implications, and alternative interpretations of the apparent restriction can undoubtedly be formulated that are equally compatible with the recruitment hypothesis.

Antisera obtained during the recovery phase have been examined in another manner by measurement of affinity by salt-precipitation methods and found normal (Dietrich and Grey, 1964). This finding would hardly be expected if recovery of responsiveness were ascribed to dissociation of antigen from an intracellular control site (since more weakly bound antigens should dissociate first) and, therefore, lends support to the recruitment hypothesis.

VI. Cellular Basis of Paralysis

Throughout the preceding discussion the assumption has been made that paralysis is caused by altering the properties of immunologically competent cells in such a way that a restricted range of reactivity is deleted. This is the mechanism that has been termed "central" failure of response, as distinct from "peripheral" failure that is brought about either by "afferent" interference with access of antigen to reactive cells, or by "efferent" interference with the antibody that is subsequently produced (Billingham et al., 1956). A peripheral mechanism of unresponsiveness is known to operate in the phenomenon of enhancement, for this can be brought about by passive transfer of antibody and, therefore, need not involve any change of cellular reactivity (Kaliss and Molomut, 1952; Kaliss, 1958). Afferent interference cannot be dismissed out of hand even in the context of paralysis by purified proteins, for this mechanism is implicit in the theory recently advanced by Eisen and Karush (1964). According to this theory, paralysis results from antigen reaching reactive cells in the form of antigen$_2$–antibody$_1$ complexes. This type of theory stands or falls according to the evidence that can be introduced for a central failure of response.

We shall survey the evidence for central failure of response during paralysis and describe how the lymphocyte has been identified as the cell affected. We shall then discuss the question as to whether or not the lymphocyte reacts directly with antigen during the process of induction. Finally, we shall consider what kinds of reaction on the part of the lymphocyte are compatible with the available evidence. Previous reviews of these topics are by Chase (1959) and Gowans and McGregor (1965).

So far as concerns the Eisen-Karush theory, and others of a similar nature, it will clarify matters if an admission is made right at the beginning. Much of the argument hinges on the outcome of "adoptive" immunization (Billingham *et al.*, 1956), i.e., upon experiments in which cells taken from normal donors are tested for reactivity in the environment provided by a paralyzed host and vice versa. On the whole this procedure provides little or no evidence of peripheral failure. Yet it can still be argued that the transferred cells themselves generate enough antibody (or fail to do so) to interfere with the reception of antigen, in which case the Eisen-Karush mechanism might still operate. All that the evidence from cell transfers can do is to impose stringent restrictions on thoeries that involve circulating antibodies rather than cells.

A. CELL TRANSFERS: EVIDENCE OF CENTRAL FAILURE OF RESPONSE

Adoptive immunization was first applied to an analysis of paralysis by Chase (1949) and by Battisto and Chase (1955, 1963) in work on unresponsiveness to allergens induced by gastric feeding. Transfer of lymph node or peritoneal exudate cells from sensitized guinea pigs confers upon noninbred recipients a transient sensitivity, and the transfer works equally well in normal and unresponsive hosts. Cells can also be transferred from unresponsive donors that have been subjected to what would normally be an effective sensitizing procedure, and these do not confer sensitivity. From these experiments it can be inferred that there is nothing to prevent the manifestation of responsiveness in the animal that has been paralyzed in this way. The final proof of central failure of response in the system is still missing, for attempts to abolish unresponsiveness by transfer of normal cells have failed, even when donors and host come from the same inbred strain (Chase, 1963).

Tolerance of transplantation antigens was the first type of paralysis to which this kind of analysis was applied in full (Billingham *et al.*, 1956). Lymph node cells were transferred from normal and from immunized donor mice into syngeneic hosts; since the homograft reaction which would otherwise have destroyed allogeneic cells was avoided (Mitchison and Dube, 1955), the belated but effective reaction on the part of the cells derived from normal donors could be examined at leisure. Because these cells are able to perform an immune response, it follows that the afferent as well as the efferent pathway to the reactive cells is open, and hence that the lesion must lie in the cells themselves (subject to the quantitative limits to this line of reasoning that have already been mentioned). Lymph node and spleen cells ought, therefore, to remain unresponsive after transfer from a tolerant donor into an irradiated recipient—a prediction that has been verified by Argyris (1962).

The same form of analysis was applied to protein paralysis by Weigle and Dixon (1959). Both a primary and a secondary response to BGG could be performed by lymph node cells transferred into paralyzed outbred rabbits. The response was detected by following the clearance of the antigen from the circulation. The primary response took place within a week after transfer, thus largely escaping the homograft reaction to which the transferred cells subsequently succumbed, although part of the erratic quality of the response could be ascribed to this reaction. In inbred mice, fairly large amounts of antibody can be produced over long periods by normal spleen cells transferred into paralyzed recipients (Mitchison, 1962c). Once again it is found that cells from paralyzed donors do not make antibody upon transfer into normal hosts (Friedman, 1962; Sercarz and Coons, 1962), and—a more stringent test—do not respond to stimulation upon transfer into irradiated hosts (Dietrich and Weigle, 1964; Stastny, 1964; Friedman, 1965a,b). The only discordant results have been obtained by Nossal and Larkin (1959) who found that cells from paralyzed donors could produce erythrocyte agglutinins after transfer into irradiated hosts; the discrepancy can be attributed to incomplete paralysis, accompanied by selection in favor of responding cells (see Section V,5).

The strongest candidate for a mechanism of unresponsiveness other than central failure is paralysis induced by pneumococcal polysaccharide. Not only is antigen retained over prolonged periods (see Section V), but it is also evidently capable of hastening the elimination of passively administered antibody (Dixon et al., 1955); these are circumstances which suggest that paralysis is the result of neutralization of antibody. Nevertheless, central failure of response has been demonstrated by adoptive immunization in paralysis by Type II (Brooke and Karnovsky, 1961) and Type I polysaccharides (Neeper and Seastone, 1963). In both cases lymph node cells from immune mice protected the paralyzed mice from challenge with viable pneumococci, and lymph node cells from paralyzed mice did not protect normal hosts.

B. THE CELL TYPE AFFECTED DURING PARALYSIS

The function of a variety of cell types might be altered in the paralyzed animal, including (1) cells of the lymphocyte series and (2) macrophages. The evidence that the small lymphocyte is the cell critically affected is now very strong for transplantation immunity and somewhat less so for serum antibody production. Details concerning transplantation immunity can be found in the admirable review of Gowans and Mc-Gregor (1965); the definitive experiment is the demonstration that cells from the thoracic duct of normal syngeneic donors can abolish tolerance of skin homografts in rats (Billingham et al., 1963; Gowans et al., 1961, 1962, 1963). McGregor and Gowans (1963) have shown that thoracic duct lymphocytes from rats paralyzed by sheep erythrocytes are incapable of restoring responsiveness to sheep erythrocytes in X-irradiated rats, although normal thoracic duct lymphocytes can do so. Thoracic duct lymphocytes have not yet been shown capable of restoring the capacity to make antibodies against protein antigens in paralyzed animals. Cell populations obtained from peripheral blood by dextran sedimentation, containing >90% lymphocytes, can restore the capacity to make antibodies to BSA and HSA (Mitchison, 1967, and unpublished data). The circumstantial evidence and the parallel with transplantation immunity leave little doubt that the small lymphocyte from the paralyzed animal in general lacks reactivity.

In accordance with this evidence of deletion of function in lymphocytes, immunoblasts and plasma cells cannot be evoked by antigenic stimulation with the inducing antigen in paralyzed animals. Sercarz and Coons (1963) could not find by the immunofluorescence method any antibody-containing cells in lymphoid organs of mice paralyzed with BSA. Cohen and Thorbecke (1964) found no histological evidence of response to challenge with BSA in paralyzed rabbits. The performance of individual plasma cells in partially paralyzed animals is a matter of some interest, since the question arises whether the response of an individual cell can be subject to partial inhibition. Occasional fluorescent cells were found by Sercarz and Coons (1963) in partially paralyzed animals, and these were as brightly stained as normal. An all-or-nothing effect was also found by Hašek et al. (1965) in their study of partially paralyzed rats by the hemolytic plaque method.

The macrophages of paralyzed animals have generally been found to deal with antigen in a normal manner when subjected to a variety of tests of immunological function. This applies, to start with, in tests of the immunogenic activity of protein antigens retained in macrophages. When this capacity is tested in the secondary response, by transferring primed cells into irradiated syngeneic hosts suspected of possessing retained antigen, BSA appears to be retained and degraded by normal and paralyzed mice in the same manner (Mitchison, 1965a). The same is true of the primary response when this is tested by transferring peritoneal macrophages from BSA-injected normal or paralyzed donors into normal mice (Mitchison, 1967).

No abnormality in macrophage behavior has been detected by the use of labeled antigen, except for what can be ascribed to concomitant immunization. Determination of radioactivity of organs in bulk indicates that the loss of ^{131}I label is normal after injection of BSA into paralyzed mice (Mitchison, 1965a). Nossal and Ada (1964) reported, however, that in rats paralyzed by flagellin, ^{125}I-labeled flagella became localized in parafollicular macrophages of draining lymph nodes just as in immune animals. This unexpected finding was modified subsequently (Ada et al., 1965) and has not been confirmed with other proteins in rabbits (Humphrey and Frank, 1967). The careful study made by Humphrey and Frank indicates that localization of HSA and hemocyanin takes place in the medullary macrophages of paralyzed animals, just as it does to start with in normal controls. In the normal animal this is followed later by the appearance of antigen in follicles, apparently because the parafollicular macrophages have the special property of localizing antibody and antigen–antibody complexes (Ada and Lang, 1966; Balfour and Humphrey, 1967).

C. Induction in Lymphocytes

Granted that the small lymphocyte has been identified as the cell ultimately affected by paralysis, the question arises whether antigen paralyzes by interaction with this cell directly, or through the mediation of the macrophage or some other cell type. A case for direct access can be made out just on the basis of the constancy of antigen dosage thresholds that was described in Table II. The same dosage of BSA, for example, is required to paralyze normal mice (low-dosage zone), newborn mice, and irradiated mice and to maintain paralysis in mice that were previously paralyzed. Under the same schedule of administration, similar doses of lysozyme are required to paralyze newborn or irradiated mice, and in irradiated mice similar requirements have also been found for diphtheria toxoid and ribonuclease. When allowance is made for body weight, even such an unusual protein as flagellin, administered to rats, appears to fall into the same pattern, and examples of similar dose requirements in other species are cited in Table II. In marked contrast to this constancy, the minimum immunizing dose varies greatly between antigens, by a factor of at least 10^4 (see Section II). From this contrast two conclusions can be drawn (*a*) the interaction between antigen and cell that leads to paralysis must be a simple one, and (*b*) a step of complexity is added for immunization to result. These conclusions accord with the hypothesis that direct access of antigen to lymphocytes induces paralysis, whereas indirect access via macrophages induces immunity.

This hypothesis was first formulated by Frei *et al.* (1965) on the basis of their work on *in vivo*-screened BSA, cited in Section II,C. It accounts satisfactorily for the capacity of high-speed centrifugation to remove immunogenic fractions from BGG (Dresser, 1962b) that has been cited; it accounts also, as Frei *et al.* point out, for the abnormally poor immunogenic quality of antigen injected into the portal circulation (Battisto and Miller, 1962) and, hence, by implication, for the special role of gastric feeding in unresponsiveness of the Chase type (Section II,B). An hypothesis of essentially the same nature was advanced by Mitchell and Nossal (1966).

The macrophage-direct access hypothesis is needed also to account for the phenomenon of overshoot during recovery that is referred to above (Section IV). If we (*1*) abandon the idea that the antigen concentration in a cell decides whether that cell is paralyzed or immune and (*2*) find direct evidence of antigen localized in macrophages at the time of overshoot, then we may conclude that antigen in this location can immunize but lacks the capacity to paralyze.

No attempt will be made here to evaluate other evidence for participation of the macrophage in immunization, nor will suggestions be offered about the hypothetical "processing" step. It is of interest, nevertheless, to note that considerations of paralysis impose both an upper and a lower limit on the complexity of the processing step. The following two considerations set the lower limit. (*1*) The role of the macrophage can hardly be simply that of concentrating antigen in order to obtain a local concentration high enough to stimulate lymphocytes, since the argument so far has run that uptake of antigen by macrophages enhances paralysis differentially. (2) Two properties are found associated in pneumococcal polysaccharides and poly-D-amino acids—slow intracellular degradation in macrophages and high potency as inducers of paralysis (Section II). This association can be accounted for most easily by assuming that a metabolic step is involved in normal macrophage processing and that resistance to the step leaves antigen that has been taken up by macrophages in a paralysis-inducing rather than immunogenic form. We assume, therefore, that antigen can be retained in or on macrophages in two different states. An analogy can be found in the recent work of Gallily and Feldman (1967) on macrophages harvested from irradiated animals: these cells can take up a bacterial antigen but not, apparently, process it into an immunogenic form. Thus, irradiated macrophages appear to handle normal antigens in the same way that normal macrophages handle pneumococcal polysaccharides and poly-D-amino acids.

The upper limit to the complexity of processing comes from a consideration of redundancy. Lymphocytes are assumed to interact directly with antigen during the induction of paralysis; the lymphocyte must, therefore, bear a special receptor that "recognizes" antigen (the conclusion holds good irrespective of any special hypothesis of receptor structure). It is, therefore, unlikely that macrophages also bear specific receptors; their role is thus more likely to be one of presenting antigen in a suitable form rather than of recognizing antigen and then transmitting to the lymphocytes a message which lacks antigenic structure.

An unequivocal demonstration of paralysis by direct action of antigen on lymphocytes is lacking. Exposure of spleen or lymph node cells, or peripheral leukocytes to purified protein antigens in vitro yields negative or variable results (R. T. Smith and J. Thorbecke, personal communications; Mitchison, 1967). Britton and Möller (1966) exposed lymphoid cells in vitro to an Escherichia coli lipopolysaccharide and found that the cells proved unresponsive to stimulation with the antigen upon transplantation into irradiated hosts. One might reasonably attribute the discrepancy with the purified protein results to the high uptake (2%) of the lipopolysaccharide; induction of paralysis may well, therefore, have taken place after the transplantation in vivo.

Cells subjected to pulse exposure to BSA or HSA in vivo have been shown to become unresponsive by means of transfer experiments (Mitchison, 1967). The exposure is given in one mouse, and the cells are then washed and transferred into another mouse (rendered unresponsive by irradiation or paralysis) for test. Exposure for over 10 minutes is required, but within 2 hours peripheral leukocytes (>90% lymphocytes) have undergone the interaction necessary for paralysis. Timing of this sort is entirely compatible with the direct-access hypothesis; similar timing has been observed in another type of lymphocyte-induction system, transformation of lymphocytes by allotypic antisera (Sell and Gell, 1965). The occurrence of a significant latent period indicates that something more than simple binding of antigen to a univalent surface receptor is involved; perhaps the formation of an antigen–receptor complex is required.

These interaction times were measured with large doses (>10 mg.) of antigen. At much lower doses (0.1–1.0 μg BSA), the rate of paralysis induction falls; responsiveness declines exponentially, with a half-time of the order of a month (Mitchison, 1967). This is a situation in which lymphocytes might be expected to undergo induction preferentially in some restricted anatomical location or physiological state. A case for the thymus as the preferred location has been made by Isakovic et al. (1965) on the basis of specific unresponsiveness found after the transplantation into irradiated rats of thymic grafts from paralyzed donors.

D. THE REACTION OF THE LYMPHOCYTE

About events in the lymphocyte subsequent to interaction with antigen we know very little in a positive sense. One question which can be tentatively answered is whether paralysis is brought about by an undetected response that differs only in minor features from a normal response. Thus Gorman and Chandler (1964) postulate a proliferative but nonproductive response that selectively suppresses the normal immune response by competition; another suggestion is that a response takes place in which antibody is produced of too low an affinity to be detected by normal methods (H. N. Eisen and L. Steiner, personal communication). In the context of high-zone paralysis, at least, this idea is not compatible with the information available concerning cellular proliferation. In diffusion chambers, overdosage with antigen lowers not only antibody titer but also synthesis of deoxyribonucleic acid, ribonucleic acid, and protein, as judged from uptake studies (Makinodan et al., 1965b); the same is true of deoxyribonucleic acid synthesis in adoptive transfers (Mäkelä and Mitchison, 1965b). The response involved in paralysis, therefore, appears not to involve proliferation.

The evidence available from studies of recovery suggests that lymphocytes are either killed or inactivated irreversibly (Section V). Irreversible inactivation might be conceived as the blocking of a receptor by combination with antigen. To pursue this idea a litttle further, one might enquire whether the known energies of binding between antigen and antibody are strong enough to account for the degree of irreversibility that has been observed. Affinities in the range of 8 to 10 kcal./mole are clearly not enough to account for the known duration of paralysis of months or even years, if the binding is simple and reversible. Introducing *ad hoc* coupling or secondary covalent linkage does not make the idea much more attractive. Death of lymphocytes and the implied absence of paralyzed cells seems an altogether simpler hypothesis; it has the merit of being easily disproved by demonstrating recovery from paralysis on the part of an isolated or transferred cellular population.

The death hypothesis (Talmage, 1957; Talmage and Pearlman, 1963; Burnet, 1959; Lederberg, 1959; Mitchell and Nossal, 1966) carries an implication for immunological theory. Since paralysis is specific, death must take place in such a manner as to leave reactivity to other antigens unaffected; a single lymphocyte must, therefore, be assumed capable of reacting against only a single antigen (a "clonal" hypothesis). Death might supervene simply as a consequence of the fixation of complement to antigen bound to molecules of receptor antibody, and part, at least, of the role of the macrophage could then be to shield the receptor from complement at the moment of binding (Nossal, 1965, and personal communication). Alternatively, death might supervene through terminal differentiation (see Section II,C) in which an abortive response takes place without cellular proliferation.

VII. Summary

Immunological paralysis is here defined as a central failure of responsiveness, brought about by exposure to antigen, in which immunologically competent cells become unable to initiate synthesis of a restricted range of antibodies. Purified proteins and carbohydrate antigens gradually induce this condition, provided that they can be maintained in the body at a concentration of over 10^{-8} to 10^{-10} M without provoking an immune response. The wide dose range refers to uncertainties in assigning a threshold rather than to variation between antigens or in the immunological status of the recipient, which is slight. Once immunity sets in, the dose of antigen required to paralyze increases, normally by a factor of the order of 10^4. This can be ascribed to competition between antibody and receptors, or, in other words, to antibody blocking access to the receptors of immunologically competent cells; the excess of antigen is required to overcome the block. We cannot dismiss the evidence for an additional mechanism whereby the response suffers transient inhibition even after the cellular receptor has been acted upon by antigen.

Agents and conditions that appear to alter the ease with which paralysis can be induced, such as age, irradiation, and adjuvants, probably do so via their effect on the mechanism of immunization; an alteration in the sensitivity of the immunologically competent cell to paralysis has not been conclusively demonstrated.

In respect of specificity, the paralysis mechanism appears to recognize wide areas of antigenic surface. Nevertheless the evidence available is compatible with the hypothesis of accurate sample receptors, i.e., that the immunologically competent cell bears as its receptor an accurate sample of the antibody which it can go on to make.

In studies of recovery from paralysis a clear case of return to competence on the part of individual cells has yet to be demonstrated. The principal means of recovery is recruitment of new competent cells via a thymus-dependent mechanism. In aged and in thymectomized animals, where recruitment occurs slowly or not at all, paralysis becomes long-lasting.

The competent cell that is affected appears to be the lymphocyte. Macrophages in or from paralyzed animals function normally, but lymphocytes display altered reactivity. There is some evidence to suggest that antigen acts directly upon lymphocytes in the induction of paralysis.

Acknowledgments

We wish to thank Miss S. Carswell and Dr. Ann Dresser for their help in the preparation of the manuscript and bibliography.

References

Ada, G. L., and Lang, P. G. (1966). *Immunology* **10**, 431.
Ada, G. L., Nossal, G. J. V., and Pye, J. (1965). *Australian J. Exptl. Biol. Med. Sci.* **43**, 337.
Adler, F. L. (1959). *In* "Mechanism of Hypersensitivity" (J. H. Shaffer, G. A. LoGrippo, and M. W. Chase, eds.), Chapt. 34. Little, Brown, Boston, Massachusetts.
Adler, F. L. (1964). *Progr. Allergy* **8**, 41.
Anderson, J. M., and Benirschke, K. (1964). *Brit. Med. J.* **i**, 1534.
Andreasen, E., and Ottesen, J. (1945). *Acta Physiol. Scand.* **10**, 258.
Argyris, B. F. (1962). *J. Immunol.* **90**, 29.
Asherson, G. L. (1966). *Immunology* **10**, 179.
Asherson, G. L., and Stone, S. H. (1965). *Immunology* **9**, 205.
Askonas, B. A., White, R. G., and Wilkinson, P. C. (1965). *Immunochemistry* **2**, 329.
Austin, C. M., and Nossal, G. J. V. (1966). *Australian J. Exptl. Biol. Med. Sci.* **44**, 341.
Azar, M. M. (1966). *J. Immunol.* **97**, 445.

Bainbridge, D. R., and Gowland, G. (1966a). *Nature* 209, 624.
Bainbridge, D. R., and Gowland, G. (1966b). *Ann. N.Y. Acad. Sci.* 129, 257.
Balfour, B. M., and Humphrey, J. H. (1967). *In* "Germinal Centers of Immune Responses" (H. Cottier, N. Odartchenko, R. Schindler, and C. C. Congdon, eds.), p. 80. Springer, Berlin.
Barth, W. F., McLaughlin, C. L., and Fahey, J. L. (1965). *J. Immunol.* 95, 781.
Bathson, B. A., Baer, H., and Shaffer, M. F. (1963). *J. Immunol.* 90, 121.
Battisto, J. R., and Bloom, B. R. (1966a). *Federation Proc.* 25, 152.
Battisto, J. R., and Bloom, B. R. (1966b). *Nature* 212, 156.
Battisto, J. R., and Chase, M. W. (1955). *Federation Proc.* 14, 456.
Battisto, J. R., and Chase, M. W. (1963). *J. Exptl. Med.* 118, 1021.
Battisto, J. R., and Chase, M. W. (1965). *J. Exptl. Med.* 121, 591.
Battisto, J. R., and Miller, J. (1962). *Proc. Soc. Exptl. Biol. Med.* 111, 111.
Benacerraf, B., Ovary, Z., Bloch, K. J., and Franklin, E. C. (1963). *J. Exptl. Med.* 117, 937.
Billingham, R. E., Brent, L., and Medawar, P. B. (1953). *Nature* 172, 603.
Billingham, R. E., Brent, L., and Medawar, P. B. (1956). *Phil. Trans. Roy. Soc. London* B239, 357.
Billingham, R. E., Silvers, W. K., and Wilson, D. B. (1963). *J. Exptl. Med.* 118, 397.
Billingham, R. E., Silvers, W. K., and Wilson, D. B. (1965). *Proc. Roy. Soc. (London)* B163, 61.
Borek, F., Stupp, Y., and Sela, M. (1965). *Science* 150, 1177.
Borel, Y., Fauconnet, M., and Miescher, P. A. (1966). *J. Exptl. Med.* 123, 585.
Boyden, S. V., and Sorkin, E. (1962). *Immunology* 5, 370.
Brent, L., and Courtenay, T. H. (1962). *Mech. Immunol. Tolerance, Proc. Symp., Liblice, Czech., 1961.* pp. 113–121.
Brent, L., and Gowland, G. (1962). *Nature* 196, 1298.
Brent, L., and Gowland, G. (1963). *Transplantation* 1, 372.
Breyere, E. J. (1964). *Med. Ann. District Columbia* 33, 93.
Breyere, E. J., and Barrett, M. K. (1960). *J. Natl. Cancer Inst.* 24, 699.
Britton, S., and Möller, G. (1966). *In* "Genetic Variations in Somatic Cells" (J. Klein, M. Vojtíšková, and V. Zelený, eds.), pp. 213–218. Czech. Acad. Sci., Prague.
Brooke, M. S. (1964). *Nature* 204, 1319.
Brooke, M. S. (1965). *Transplantation* 3, 478.
Brooke, M. S. (1966). *Transplantation* 4, 1.
Brooke, M. S., and Karnovsky, M. J. (1961). *J. Immunol.* 87, 205.
Brownstone, A., Mitchison, N. A., and Pitt-Rivers, R. (1966). *Immunology* 10, 481.
Burnet, F. M. (1959). "The Clonal Selection Theory of Acquired Immunity." Cambridge Univ. Press, London and New York.
Burnet, F. M., and Fenner, F. (1949). "The Production of Antibodies," 2nd Ed. Macmillan, New York.
Bussard, A. E. (1962). *Mech. Immunol. Tolerance, Proc. Symp., Liblice, Czech., 1961* pp. 85–94.
Calabresi, P., and Welch, A. D. (1965). *In* "The Pharmacological Basis of Therapeutics" (L. S. Goodman and A. Gilman, eds.), Chapt. 62, pp. 1345–1392. Macmillan, New York.
Černý, J., Ivanyi, J., Madar, J., and Hraba, T. (1965). *Folia Biol. (Prague)* 11, 402.
Chase, M. W. (1949). *Proc. 49th Gen. Meeting, Soc. Am. Bacteriologists* p. 75 (abstr.).

Chase, M. W. (1959). *Ann. Rev. Microbiol.* **13**, 349.

Chase, M. W. (1963). *In* "La Tolérance Acquise et la Tolérance Naturelle à l'Égard de Substances Antigéniques Définies" (A. Bussard, ed.), pp. 139–160. C.N.R.S., Paris.

Chase, M. W., and Battisto, J. R. (1959). *In* "Mechanisms of Hypersensitivity" (J. H. Shaffer, G. A. LoGrippo, and M. W. Chase, eds.), p. 507. Churchill, London.

Chou, C.-T., Dubiski, S., and Cinader, B. (1966). *Nature* **211**, 34.

Cinader, B., and Dubert, J. M. (1955). *Brit. J. Exptl. Pathol.* **36**, 515.

Cinader, B., and Dubert, J. M. (1956). *Proc. Roy. Soc. (London)* **B146**, 18.

Claman, H. N. (1963). *J. Immunol.* **91**, 833.

Claman, H. N., and Bronsky, E. A. (1966). *J. Allergy* **38**, 208.

Claman, H. N., and McDonald, W. (1964). *Nature* **202**, 712.

Claman, H. N., and Talmage, D. W. (1963). *Science* **141**, 1193.

Coe, J. E. (1966). *J. Immunol.* **96**, 744.

Coe, J. E., and Salvin, S. B. (1963). *J. Exptl. Med.* **117**, 401.

Coe, J. E., and Salvin, S. B. (1964). *J. Immunol.* **93**, 495.

Cohen, M. W., and Thorbecke, G. J. (1963). *Proc. Soc. Exptl. Biol. Med.* **112**, 10.

Cohen, M. W., and Thorbecke, G. J. (1964). *J. Immunol.* **93**, 629.

Cohen, S. (1965). *Immunology* **8**, 1.

Coons, A. H. (1963). *In* "La Tolérance Acquise et la Tolérance Naturelle à l'Égard de Substances Antigéniques Définies" (A. Bussard, ed.), pp. 121–132. C.R.N.S., Paris.

Crowle, A. J. (1963). *J. Allergy* **34**, 504.

Crowle, A. J., and Hu, C. C. (1966). *Clin. Exptl. Immunol.* **1**, 323.

Cruchaud, A. (1966). *J. Immunol.* **96**, 832.

Curtain, C. C. (1959). *Brit. J. Exptl. Pathol.* **40**, 255.

Day, J. H., and Farr, R. S. (1966). *Immunology* **11**, 571.

Deichmiller, M. P., and Dixon, F. J. (1960). *J. Gen. Physiol.* **43**, 1047.

Denhardt, D. T., and Owen, R. D. (1960). *Transplant. Bull.* **7**, 394.

Dietrich, F. M. (1966). *J. Immunol.* **97**, 216.

Dietrich, F. M., and Grey, H. M. (1964). *Nature* **201**, 1236.

Dietrich, F. M., and Weigle, W. O. (1963). *J. Exptl. Med.* **117**, 621.

Dietrich, F. M., and Weigle, W. O. (1964). *J. Immunol.* **92**, 167.

Dixon, F. J., and McConahey, P. J. (1963). *J. Exptl. Med.* **117**, 833.

Dixon, F. J., and Maurer, P. H. (1955). *J. Immunol.* **74**, 418.

Dixon, F. J., Maurer, P. H., and Weigle, W. O. (1955). *J. Immunol.* **74**, 188.

Dorner, M. M., and Uhr, J. W. (1964). *J. Exptl. Med.* **120**, 435.

Dresser, D. W. (1961a). *Nature* **191**, 1169.

Dresser, D. W. (1961b). *Immunology* **4**, 13.

Dresser, D. W. (1962a). *Immunology* **5**, 161.

Dresser, D. W. (1962b). *Immunology* **5**, 378.

Dresser, D. W. (1963). *Immunology* **6**, 345.

Dresser, D. W. (1965). *Immunology* **9**, 261.

Dresser, D. W., and Gowland, G. (1964). *Nature* **203**, 733.

Dvorak, H. F., and Flax, M. H. (1966). *J. Immunol.* **96**, 546.

Dvorak, H. F., Billote, J. B., McCarthy, J. S., and Flax, M. H. (1965). *J. Immunol.* **94**, 966.

Dvorak, H. F., Billote, J. B., McCarthy, J. S., and Flax, M. H. (1966). *J. Immunol.* **97**, 106.

Eichwald, E. J. (1963). *Advan. Biol. Med. Phys.* **9**, 93.

Eisen, H., and Karush, F.(1964). *Nature* **202**, 677.

Fahey, J. L., Wunderlich, J., and Mishell, R. (1964). *J. Exptl. Med.* **120**, 223.

Fefer, A., and Nossal, G. J. V. (1962). *Transplant. Bull.* **29**, 445.

Felton, L. D. (1949). *J. Immunol.* **61**, 107.

Felton, L. D., and Ottinger, B. (1942). *J. Bacteriol.* **43**, 94.

Felton, L. D., Kauffmann, G., Prescott, B., and Ottinger, B. (1955a). *J. Immunol.* **74**, 17.

Felton, L. D., Prescott, B., Kauffmann, G., and Ottinger, B. (1955b). *J. Immunol.* **74**, 205.

Finkelstein, M. S., and Uhr, J. W. (1964). *Science* **146**, 67.

Ford, C. E., and Micklem, H. S. (1963). *Lancet* **i**, 359.

Frei, F. C. (1964). *Helv. Physiol. Pharmacol. Acta* **22**, 124.

Frei, F. C., Benacerraf, B., and Thorbecke, G. J. (1965). *Proc. Natl. Acad. Sci. U.S.* **53**, 20.

Freund, J. (1930). *J. Immunol.* **18**, 313.

Frey, J. R., Geleick, H., and de Weck, A. (1964). *Science* **144**, 853.

Friedman, H. (1962). *J. Immunol.* **89**, 257.

Friedman, H. (1964). *J. Immunol.* **92**, 201.

Friedman, H. (1965a). *J. Immunol.* **94**, 352.

Friedman, H. (1965b). *Transplantation* **3**, 465.

Friedman, H. (1965c). *J. Immunol.* **94**, 205.

Friedman, H. (1966). *J. Bacteriol.* **92**, 820.

Gaines, S., Currie, J. A., and Tully, J. G. (1966). *Ann. Inst. Pasteur* **110**, 60.

Gallily, R., and Feldman, M. (1967). *Immunology* **12**, 197.

Gell, P. G. H., and Kelus, A. S. (1966). *Nature* **211**, 766.

Gill, T. J. (1965). *J. Immunol.* **95**, 542.

Gill, T. J., Papermaster, D. S., and Mowbray, J. F. (1965). *J. Immunol.* **95**, 794.

Glenny, A. T., and Hopkins, B. E. (1924). *J. Hyg.* **22**, 208.

Good, R. A., Kelly, W. D., Martinez, C., Pollara, B., Holmes, B. M., McKneally, M. F., and Gabrielsen, A. E. (1966). *In* "Immunopathology" (P. Graber and P. A. Miescher, eds.), pp. 145–156. Benno Schwabe, Basel.

Gorman, J. G., and Chandler, J. G. (1964). *Blood* **23**, 117.

Gowans, J. L., and McGregor, D. D. (1965). *Progr. Allergy* **9**, 1.

Gowans, J. L., Gesner, B. M., and McGregor, D. D. (1961). *Ciba Found. Study Group* **10**, 32.

Gowans, J. L., McGregor, D. D., Cowen, D. M., and Ford, C. E. (1962). *Nature* **196**, 651.

Gowans, J. L., McGregor, D. D., and Cowen, D. M. (1963). *Ciba Found. Study Group* **16**, 20.

Gowland, G. (1965). *Brit. Med. Bull.* **21**, 123.

Gowland, G., Hobbs, G., and Byers, H. D. (1965). *J. Pathol. Bacteriol.* **90**, 443.

Hanan, R., and Oyama, J. (1954). *J. Immunol.* **73**, 49.

Hašek, M. (1953). *Czechoslovak Biol.* **2**, 265.

Hašek, M., Lengerova, A., and Hraba, T. (1961). *Advan. Immunol.* **1**, 1.

Hašek, M., Hraba, T., and Madar, J. (1965). *Folia Biol.* (*Prague*) **11**, 318.

Howard, J. G., and Michie, D. (1962). *Transplant. Bull.* **29**, 1.

Humphrey, J. H. (1961). *Immunology* 4, 380.

Humphrey, J. H. (1964a). *Immunology* 7, 449.

Humphrey, J. H. (1964b). *Immunology* 7, 462.

Humphrey, J. H. (1965). *In* "Immunologic Diseases" (M. Samter, ed.), pp. 100–108. Little, Brown, Boston, Massachusetts.

Humphrey, J. H., and Frank, M. M. (1967). *Immunology* 13, 87.

Humphrey, J. H., and Turk, J. L. (1961). *Immunology* 4, 301.

Isakovic, K., Smith, S. B., and Waksman, B. H. (1965). *Science* 148, 1333.

Ivanyi, J., Hraba, T., and Černý, J. (1964). *Folia Biol. (Prague)* 10, 198.

Janeway, C. A., and Sela, M. (1967). *Immunology* 13, 29.

Jensen, E., and Simonsen, M. (1963). *Ann. N.Y. Acad. Sci.* 99, 657.

Jones, V. E., and Leskowitz, S. (1965). *J. Exptl. Med.* 122, 505.

Kabat, E. A., and Meyer, M. M. (1961). "Experimental Immunochemistry," 2nd Ed. Thomas, Springfield, Illinois.

Kaliss, N. (1958). *Cancer Res.* 18, 992.

Kaliss, N., and Molomut, N. (1952). *Cancer Res.* 12, 110.

Kamrin, B. B. (1959). *Proc. Soc. Exptl. Biol. Med.* 100, 58.

Kamrin, B. B. (1966). *Science* 153, 1261.

Kaplan, M. H., Coons, A. H., and Deane, H. W. (1950). *J. Exptl. Med.* 91, 15.

Krenova-Peclova, D., Kral, J., Baborovska, J., and Kren, V. (1963). *Folia Biol. (Prague)* 9, 258.

Landsteiner, E. K. (1936). "The Specificity of Serological Reactions." Thomas, Springfield, Illinois.

Lederberg, J. (1959). *Science* 129, 1649.

Leskowitz, S. (1967). *Immunology* 13, 9.

Levey, R. H., and Medawar, P. B. (1966). *Proc. Natl. Acad. Sci. U.S.* 56, 1130.

Levey, R. H., and Medawar, P. B. (1967). *Ciba Found. Study Group,* 29.

Levey, R. H., Trainin, N., and Law, L. W. (1963). *J. Natl. Cancer Inst.* 31, 199.

Liacopoulos, P. (1961). *Compt. Rend.* 253, 751.

Liacopoulos, P., and Neveu, T. (1964). *Immunology* 7, 26.

Liacopoulos, P., and Perramant, F. (1966). *Ann. Inst. Pasteur* 110, Suppl., 161.

Liacopoulos, P., Halpern, B. N., and Perramant, F. (1962). *Nature* 195, 1112.

Linscott, W. D., and Weigle, W. O. (1965a). *J. Immunol.* 94, 430.

Linscott, W. D., and Weigle, W. O. (1965b). *J. Immunol.* 95, 546.

Lowney, E. D. (1965). *J. Immunol.* 95, 397.

McBride, R. A., and Simonsen, M. (1965). *Transplantation* 3, 140.

McGregor, D. D., and Gowans, J. L. (1963). *J. Exptl. Med.* 117, 303.

McGregor, D. D., and Gowans, J. L. (1964). *Lancet* i, 629.

McKhann, C. F. (1962). *J. Immunol.* 88, 500.

McKhann, C. F. (1964). *Transplantation* 2, 620.

Mäkelä, O., and Mitchison, N. A. (1965a). *Immunology* 8, 539.

Mäkelä, O., and Mitchison, N. A. (1965b). *Immunology* 8, 549.

Mäkelä, O., and Nossal, G. J. V. (1962). *J. Immunol.* 88, 613.

Makinodan, T., and Gengozian, N. (1960). *In* "Radiation Protection and Recovery" (A. Hollaender, ed.), pp. 316–351. Macmillan (Pergamon), New York.

Makinodan, T., Allbright, J. F., Perkins, E. H., and Nettescheim, P. (1965a). *Med. Clin. N. Am.* 49, 1569.

Makinodan, T., Hoppe, I., Sado, T., Capalbo, E. E., and Leonard, M. R. (1965b). *J. Immunol.* 95, 466.

Mariani, T., Martinez, C., Smith, J. M., and Good, R. A. (1959). *Proc. Soc. Exptl. Biol. Med.* **101**, 596.

Maurer, P. H., Lowy, R., and Kierney, C. (1963). *Science* **139**, 1061.

Medawar, P. B. (1963). *Transplantation* **1**, 21.

Metcalf, D. (1965). *Nature* **208**, 1336.

Miller, J. F. A. P. (1964). *Science* **144**, 1544.

Miller, J. F. A. P. (1965). *Nature* **208**, 1337.

Mitchell, J., and Nossal, G. J. V. (1966). *Australian J. Exptl. Biol. Med. Sci.* **44**, 211.

Mitchison, N. A. (1959). In "Colloque International sur les Problems Biologiques des Greffes," pp. 239–255. Univ. de Liège, Liège, Belgium.

Mitchison, N. A. (1961). *Brit. Med. Bull.* **17**, 102.

Mitchison, N. A. (1962a). *Immunology* **5**, 341.

Mitchison, N. A. (1962b). *Immunology* **5**, 359.

Mitchison, N. A. (1962c). *Mech. Immunol. Tolerance, Proc. Symp., Liblice, Czech., 1961* pp. 245–255.

Mitchison, N. A. (1964). *Proc. Roy. Soc. (London)* **B161**, 275.

Mitchison, N. A. (1965a). *Immunology* **9**, 129.

Mitchison, N. A. (1965b). *Proc. Intern. Symp. Immunol. Tolerance* p. 364. Locarno.

Mitchison, N. A. (1966). *Progr. Biophys. Mol. Biol.* **16**, 3.

Mitchison, N. A. (1967). In "Regulation of the Antibody Response" (B. Cinader, ed.). Thomas, Springfield, Illinois. In press.

Mitchison, N. A., and Dube, O. L. (1955). *J. Exptl. Med.* **102**, 179.

Mowbray, J. F. (1963a). *Immunology* **6**, 217.

Mowbray, J. F. (1963b). *Transplantation* **1**, 15.

Mowbray, J. F., and Hargrave, D. C. (1966). *Immunology* **11**, 413.

Mowbray, J. F., and Scholand, J. (1966). *Immunology* **11**, 421.

Muller-Eberhard, U., English, E. C., and Weigle, W. O. (1965). *Proc. Soc. Exptl. Biol. Med.* **120**, 11.

Nachtigal, D., and Feldman, M. (1964). *Immunology* **7**, 616.

Nachtigal, D., Eschel-Zussman, R., and Feldman, M. (1965). *Immunology* **9**, 543.

Neeper, C. A., and Seastone, C. V. (1963). *J. Immunol.* **91**, 374.

Nossal, G. J. V. (1965). *Proc. Intern. Symp. Immunol. Tolerance* p. 380. Locarno.

Nossal, G. J. V., and Ada, G. L. (1964). *Nature* **201**, 580.

Nossal, G. J. V., and Austin, C. M. (1966). *Australian J. Exptl. Biol. Med. Sci.* **44**, 327.

Nossal, G. J. V., and Larkin, L. (1959). *Australian J. Sci.* **22**, 168.

Nossal, G. J. V., Ada, G. L., and Austin, C. M. (1965). *J. Immunol.* **95**, 665.

Osoba, D., and Miller, J. F. A. P. (1963). *Nature* **199**, 653.

Owen, R. D. (1945). *Science* **102**, 400.

Pace, M. G., and Dresser, D. W. (1961). *Quart. J. Exptl. Physiol.* **46**, 369.

Phillips, J. M. (1966). *Immunology* **11**, 163.

Porter, J. B., and Breyere, E. J. (1964). *Transplantation* **2**, 246.

Porter, R. J. (1962). *Federation Proc.* **21**, 28 (abstr.).

Rowley, D. A., and Fitch, F. W. (1965a). *J. Exptl. Med.* **121**, 671.

Rowley, D. A., and Fitch, F. W. (1965b). *J. Exptl. Med.* **121**, 683.

Rubin, H., Fanshier, L., Cornelius, A., and Hughes, W. F. (1962). *Virology* **17**, 143.

Sahiar, K., and Schwartz, R. S. (1964). *Science* **145**, 395.

Sahiar, K., and Schwartz, R. S. (1965). *J. Immunol.* **95**, 345.

Sahiar, K., and Schwartz, R. S. (1966). *Intern. Arch. Allergy Appl. Immunol.* **29**, 52.

Schechter, I., Bauminger, S., and Sela, M. (1964a). *Biochim. Biophys. Acta* **93**, 686.

Schechter, I., Bauminger, S., Sela, M., Nachtigal, D., and Feldman, M. (1964b). *Immunochemistry* **1**, 249.

Schwartz, J., Kletter, J., and Klopstock, A. (1964). *Intern. Arch. Allergy Appl. Immunol.* **25**, 83.

Schwartz, R. S. (1966). *Federation Proc.* **25**, 165.

Schwartz, R. S., and Dameshek, W. (1963). *J. Immunol.* **90**, 703.

Seamer, J. (1965). *Arch. Ges. Virusforsch.* **15**, 169.

Sela, M., Fuchs, S., and Feldman, M. (1963). *Science* **139**, 342.

Sell, S., and Gell, P. G. H. (1965). *J. Exptl. Med.* **122**, 423.

Sercarz, E., and Coons, A. H. (1962). *Mech. Immunol. Tolerance, Proc. Symp., Liblice, Czech., 1961* pp. 73–83.

Sercarz, E., and Coons, A. H. (1963). *J. Immunol.* **90**, 478.

Shaul, D. B. (1962). *Bull. Res. Council Israel* **E10**, 45.

Shaul, D. B. (1963). *Israel J. Exptl. Med.* **11**, 18.

Silverstein, A. M., Uhr, J. W., Kraner, K. L., and Lukes, R. J. (1963). *J. Exptl. Med.* **117**, 799.

Singer, S. J. (1964). *Immunochemistry* **1**, 15.

Siskind, G. W., and Howard, J. G. (1966). *J. Exptl. Med.* **124**, 417.

Siskind, G. W., Paterson, P. Y., and Thomas, L. (1963). *J. Immunol.* **90**, 929.

Smith, R. T. (1961). *Advan. Immunol.* **1**, 67.

Smith, R. T., and Bridges, R. A. (1958). *J. Exptl. Med.* **108**, 227.

Sobey, W. R., Magrath, J. M., and Reisner, A. H. (1966). *Immunology* **11**, 511.

Staples, P. J., Gery, I., and Waksman, B. H. (1966). *J. Exptl. Med.* **124**, 127.

Stark, O. K. (1955a). *J. Immunol.* **74**, 126.

Stark, O. K. (1955b). *J. Immunol.* **74**, 130.

Stastny, P. (1964). *J. Immunol.* **92**, 626.

Sterzl, J. (1966). *Nature* **209**, 416.

Stone, W. H., and Owen, R. D. (1963). *Transplantation* **1**, 107.

Stoner, R. D., and Hale, W. M. (1962). *In* "The Effects of Ionizing Radiations and Immune Processes" (C. A. Leone, ed.), pp. 183–219. Gordon & Breach, New York.

Streilein, J. W., and Hildreth, E. A. (1966). *J. Immunol.* **96**, 1027.

Sulzberger, M. B. (1929). *Arch. Dermatol. Syphilol.* **20**, 669.

Sulzberger, M. B. (1930). *Arch. Dermatol. Syphilol.* **22**, 839.

Taliaferro, W. H. (1957). *Ann. N.Y. Acad. Sci.* **69**, 745.

Talmage, D. W. (1957). *Ann. Rev. Med.* **8**, 239.

Talmage, D. W., and Pearlman, D. S. (1963). *J. Theoret. Biol.* **5**, 321.

Taylor, R. B. (1963). *Nature* **199**, 873.

Taylor, R. B. (1964). *Immunology* **7**, 595.

Taylor, R. B. (1965). *Nature* **208**, 1334.

Tempelis, C. H. (1965). *J. Immunol.* **94**, 705.

Tempelis, C. H., Wolfe, H. R., and Mueller, A. (1958). *Brit. J. Exptl. Pathol.* **39**, 323.

Terres, G., and Hughes, W. L. (1959). *J. Immunol.* **83**, 459.

Terres, G., and Wolins, W. (1959). *J. Immunol.* **83**, 9.

Thompson, J. S., Crawford, M., Severson, D., and Russe, H. P. (1966). *Nature* **211**, 1155.

Thorbecke, G. J., Siskind, G. W., and Goldberger, N. (1961). *J. Immunol.* **87**, 147.

Turk, J. L., and Humphrey, J. H. (1961). *Immunology* **4**, 310.

Uhr, J. W. (1967). *In* "Regulation of the Antibody Response" (B. Cinader, ed.), Thomas, Springfield, Illinois. In press.

Uhr, J. W., and Baumann, J. B. (1961). *J. Exptl. Med.* **113**, 935.

Weigle, W. O. (1961). *J. Exptl. Med.* **114**, 111.

Weigle, W. O. (1962). *J. Exptl. Med.* **116**, 913.

Weigle, W. O. (1964a). *J. Immunol.* **92**, 113.

Weigle, W. O. (1964b). *Nature* **201**, 632.

Weigle, W. O. (1964c). *Immunology* **7**, 239.

Weigle, W. O. (1965a). *J. Exptl. Med.* **121**, 289.

Weigle, W. O. (1965b). *J. Immunol.* **94**, 177.

Weigle, W. O. (1966a). *Federation Proc.* **25**, 160.

Weigle, W. O. (1966b). *Intern. Arch. Allergy Appl. Immunol.* **29**, 254.

Weigle, W. O., and Dixon, F. J. (1959). *J. Immunol.* **82**, 516.

Weiss, D. W., and Main, O. (1962). *Immunology* **5**, 333.

Weissman, I. L. (1966). *Transplantation* **5**, 565.

White, R. G., Jenkins, G. C., and Wilkinson, P. C. (1963). *Intern. Arch. Allergy Appl. Immunol.* **22**, 156.

Wilkinson, P. C., and White, R. G. (1966). *Immunology* **11**, 229.

Woodruff, M. F. A., and Anderson, N. A. (1963). *Nature* **200**, 702.

Zubay, G. (1963). *Nature* **200**, 483.

In Vitro Studies of Human Reaginic Allergy

ABRAHAM G. OSLER, LAWRENCE M. LICHTENSTEIN,[1,2]
AND DAVID A. LEVY[1,3]

Department of Microbiology, The Johns Hopkins University School of Medicine, Baltimore, Maryland

I. Introduction

In many respects the human pollinoses provide appropriate and convenient models for the study of immunological diseases attributable to allergic reactions of the immediate type. In diseases such as hay fever, the specific antigens are usually known, the etiologic role of the immune event can be readily established, and several mediators of the tissue response have been identified and characterized. Despite these favorable circumstances, elucidation of the underlying immunological and cellular mechanisms has but barely begun. Witness for example the anachronistic but widespread use of the term "reagin," coined in 1908 by Citron to avoid the commitment that the serum reactant in tests for syphilis is,

[1] Work done as Fellow, United States Public Health Service Graduate Training Grant No. 2 T1-GM-624 to the Department of Medicine, The Johns Hopkins University School of Medicine.
[2] Present address: Research and Teaching Scholar of the American College of Physicians, Department of Medicine, The Johns Hopkins Hospital, Baltimore, Maryland.
[3] Present address: Department of Radiological Sciences, The Johns Hopkins University School of Hygiene and Public Health, Baltimore, Maryland.

indeed, a specific antibody. Adoption of this term in 1925 by Coca and Grove for allergic diseases of the hay fever type reflected similar doubts, probably because an adequate methodology for *in vitro* studies of human pollinoses was not available.

Many experimental procedures have, of course, been used by students of human allergic disease and these are well known (reviewed by Stanworth, 1963). Numerous serological procedures (e.g., hemagglutination, antigen binding, etc.) have been applied to this problem. Unfortunately, these suffer from the defect that the results cannot be related in a meaningful manner to the clinical status of the pollinosis patient. In contrast, skin tests have been used successfully for the detection and assay of the incriminated antigens and the antibodies formed during the natural course of the disease (Prausnitz and Küstner, 1921) or as a result of therapeutic immunization (Cooke *et al.*, 1935, 1937). Despite their great utility, the manipulative and interpretational difficulties inherent in skin test studies have proven quite formidable. Difficulties of the former type may be alleviated in part through the use of nonhuman primates for skin tests and Schultz-Dale studies (Layton *et al.*, 1963; Layton, 1965; Rose *et al.*, 1964; Arbesman *et al.*, 1964). However, studies with intact animals or their tissues are not suitable for characterization at the cellular level, of the individual reaction steps in the sequence which begins with an antigen–antibody reaction and which terminates in the release of vasoactive agents.

Within recent years several areas of research have been developed whose completion give promise that the major problems in our understanding of human pollinosis may soon be resolved. Among these are the clarification of structure–function relationships of different classes of immunoglobulins (reviewed in Bloch, 1967), purification of the offending antigens (King and Norman, 1962; King *et al.*, 1964), and the availability of biologically valid *in vitro* procedures for the study of human reaginic allergy. This review will be devoted almost exclusively to the latter and, more specifically, to the use of human leukocytes for studies of ragweed hay fever. With few exceptions, little effort will be made to relate these findings to the extensive body of literature dealing with *in vitro* studies of immediate allergic reactions based on the use of experimental animals or their tissues. These have been well summarized in two recent reviews (Mongar and Schild, 1962; Austen and Humphrey, 1963).

The choice of human leukocytes for the construction of an *in vitro* reaction system was based on several considerations. Perhaps the foremost was the desire to use a cell suspension in order to circumvent the technical limitations imposed by intact tissues or their fragments. The use of a suspension of cells serving both as carriers of the mediating immunoglobulins and as reservoirs of tissue permeability factors seemed particularly well-suited for a variety of experimental approaches. Katz and Cohen (1941) had already demonstrated that the addition of specific antigen to the blood of an allergic donor increased plasma histamine levels—an observation effectively extended by Noah and Brand (1954, 1955). The analyses of Graham and her collaborators (1955) were noteworthy in showing that almost all of the human blood histamine was contained in the polymorphonuclear cells and primarily in the basophiles. Nevertheless, subsequent experiments were still largely conducted with whole blood (VanArsdel et al., 1958; VanArsdel and Middleton, 1961; Middleton and Sherman, 1960; Middleton et al., 1960). Reports by the latter groups extended the earlier findings regarding the release of histamine by specific antigen and recorded some effects of therapeutic immunization. Among these were the experiments of Middleton and Sherman (1960) which tended to exclude the participation of the complement (C') system in the sequence leading to the immune release of histamine. The use of whole human blood in much of this work created some interpretational ambiguities since the contributions of the cellular and fluid elements could not be individually evaluated.

On the basis of these reports, a reaction system was designed for *in vitro* studies of allergic histamine release from washed human leukocytes suspended in a relatively simple, serumfree buffer. Initially, suspensions of cells from ragweed-sensitive donors were used to demarcate some of the nonimmunological parameters governing the release of histamine by ragweed pollen antigen. Methods were then devised, as detailed below, to assay both blocking and sensitizing antibody activities, in order to estimate donor cell sensitivity to ragweed pollen antigens and to characterize the effects of therapeutic immunization on the clinical course of ragweed hay fever. This report is devoted to a discussion of these subjects.

There are four principal reactants in these studies, viz., the histamine-containing cell; the antibody of pathogenetic importance (i.e., the reagin) which by "fixing" to the cell renders it reactive to antigen; the inciting antigen, which for ragweed is primarily a protein with a molecular weight of 38,000; and the "blocking" antibody, which by combining with antigen in the fluid phase can inhibit the reaction between the antigen and the cell-fixed antibody that triggers the *in vitro* allergic response. Each of these reactants can be varied independently to determine its influence on the overall reaction. Since tissue or cell sensitization marks one of the early and key steps in the immunopathological sequence, this area of study will be dealt with first. However, a preliminary statement regarding the histamine-releasing procedure may be pertinent.

II. Procedure for Estimating Histamine Release from Human Leukocytes

Complete details regarding the technical procedures may be found in the original publications (Lichtenstein and Osler, 1964, 1966a,b; Lichtenstein et al., 1966a,b). Human blood serves as the source of leukocytes which are separated according to a slight modification of the technique of Lapin et al. (1958). All surfaces that the leukocytes come into contact with are siliconized glass or plastic. Following separation from the other formed blood elements in a mixture of ethylenediaminetetraacetic acid (EDTA), glucose, dextran, and physiological saline, the leukocytes are washed by two cycles of centrifugation and resuspension in a tris-buffered diluent at 4°C. The latter contains human serum albumin (HSA), 0.3 mg./ml., as well as calcium and magnesium for reasons which are detailed below. The final suspension contains from 1 to 2×10^7 leukocytes per milliliter, an approximately equivalent number of erythrocytes, and about four platelets per white cell. Since human red blood cells contain no histamine, and platelets but insignificant amounts, their inclusion in the reaction mixtures is of little, if any, importance. As will be shown below, leukocytes prepared in this fashion still carry a variety of plasma constituents including immunoglobulins and C' components.

The standardized leukocyte suspension is incubated with selected levels of antigen or other reagents, and the reaction is terminated by centrifugation in the cold. The amount of histamine liberated into the fluid phase of the reaction mixtures is estimated by the fluorometric procedure of Shore et al. (1959), as modified by Kremzner and Wilson (1961) and Lichtenstein and Osler (1964). The histamine content in an aliquot of the cell suspension, determined after disruption of the cells with perchloric acid, is used as a reference value for calculating the portion of this amine liberated under the specific experimental conditions.

III. Passive Sensitization, in Vitro, of Human Leukocytes

In view of the reports by Middleton (1960) and VanArsdel and Sells (1963) that in vitro sensitization could be achieved, a systematic study was undertaken of the interaction between serum from ragweed-sensitive subjects and donor leukocytes screened for their lack of responsiveness to ragweed pollen antigen E. This part of the investigation had two objectives. These were the development of valid *in vitro* procedures for studies of the mechanism of antibody fixation and for the assay of human reaginic or skin-sensitizing antibodies.

A. GENERAL COMMENTS

Passive sensitization of ragweed-insensitive leukocytes is achieved by incubating these cells with serum from an allergic subject at 37°C., as detailed below. The leukocytes are then separated from the fluid

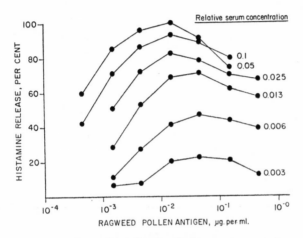

FIG. 1. Histamine release from human leukocytes sensitized passively with different amounts of serum from a ragweed-sensitive individual.

phase constituents by two cycles of centrifugation and resuspension in tris buffer, and ragweed antigen is added (Levy and Osler, 1966). The cellular release of histamine is estimated as described by Lichtenstein and Osler (1964). The extent of sensitization is measured indirectly, i.e., in terms of histamine release. As shown in Fig. 1, this response is a function of the serum concentration. It should be noted that the sensitizing potency is only an activity measurement and provides no information as to the actual quantity or character of the participating immunoglobulins.

B. PARAMETERS OF LEUKOCYTE SENSITIZATION

In vitro sensitization of human leukocytes conforms to physiological conditions with respect to pH and temperature (Levy and Osler, 1966). Optimal pH values center about 7.4 and the rate of sensitization increases rapidly as the reaction temperature is raised from 4 to 37°C. The relative rates of sensitization at 4, 20, and 37°C., as determined with one serum (Fig. 2), are 0.08, 0.2 and 1.0. Estimation of the activation energy based on these data yielded a value of about 9000 cal./mole which falls within the range of many other biological processes. The sensitization of guinea pig ileum (Halpern *et al.*, 1958; Nielsen *et al.*, 1959) or lung fragments (Mongar and Schild, 1957; Brocklehurst *et al.*, 1961) also proceeds more rapidly at 37°C. than at lower temperatures.

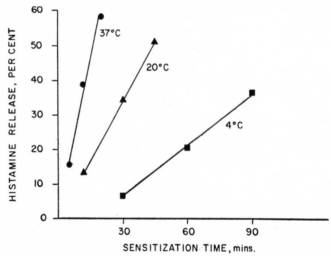

FIG. 2. Effect of temperature on passive sensitization of human leukocytes *in vitro*. *Exp. 050365*. Cells, FrAt; histamine available, 0.31 μg. per 1.5 × 10⁷ white blood cells (wbc); serum, KeTu, 111363, 1:20; ragweed pollen antigen, 1.5 × 10⁻² μg./ml.

1. Enhancing Effect of EDTA

Experiments such as summarized in Table I led to the unexpected finding that chelation of divalent cations promotes the *in vitro* fixation of ragweed-sensitizing antibodies to human leukocytes. This observation was confirmed in numerous other experiments, as shown in Figs. 3 and 4, which also served to identify the inhibitory cation as calcium (cf., however, Mongar and Schild, 1960; Humphrey *et al.*, 1963; VanArsdel and Sells, 1963). For the experiment depicted in Fig. 3, leukocytes from two normal donors were treated with sera from allergic individuals in the presence of 0.004 M EDTA. These cells were washed and after resuspension and antigen addition, they released about 70% of their histamine.

TABLE I

Passive Sensitization of Human Leukocytes *in Vitro;* Enhancement of
Histamine Release When Ethylenediaminetetraacetic Acid Is
Present during Sensitization[a]

EDTA present during:			Histamine release (%)
Cell separation (10 mM)	Cell washing (5 mM)	Sensitization with serum (5 mM)	
+	0	0	24
+	+	0	30
+	0	+	53
+	+	+	51

[a] Exp. 030264. Cells, FrAt; histamine available, 0.27 μg. per 1.1×10^7 white blood cells. Serum, ViMa, 022864, 1:5. Ragweed pollen antigen, 4.5×10^{-3} μg./ml.

Replicate cell suspensions, also incubated with this quantity of EDTA and increasing amounts of calcium, were washed and then reacted with antigen. Calcium and, to a lesser extent, magnesium inhibit the response as shown in Fig. 3. Virtually complete suppression is noted with $4.1 \times 10^{-3} M$ Ca^{2+}. The data in Fig. 4 show a similar effect of calcium and magnesium with leukocytes sensitized *in vivo,* i.e., taken from ragweed-sensitive donors. It thus emerges that calcium exerts a dual role in the allergic reaction of human leukocytes. This cation is essential for the responsiveness of a reactive cell to antigen, yet preincubation of the sensitized cell with calcium renders the leukocyte less responsive. Conceivably, the action of divalent cations may be related to an alteration of charge on the cell surface as suggested by Wilkins *et al.* (1962). They reported a marked reduction in the electrophoretic mobility of sheep leukocytes bathed in $0.01 M$ calcium. It remains to be established whether calcium acts by reducing the capacity of the cells to bind or retain the sensitizing immunoglobulins or at an undisclosed but later step which is only manifest in the release process.

2. Synergistic Action of Heparin and EDTA

The increased responsiveness of human leukocytes following sensitization in the presence of EDTA suggested that the activity of this compound might be related to its anticoagulant properties. Heparin was, therefore, employed as an additional compound with similar properties but operating at a step other than the chelation of divalent cations. The results of experiments such as those summarized in Fig. 5 leave little

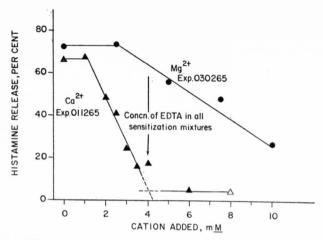

FIG. 3. Effect of calcium or magnesium on passive sensitization of human leukocytes *in vitro*. *Exp. 011265.* Cells, RoTh; histamine available, 0.13 μg. per 1.7×10^7 wbc; serum, DiGu, 093064, 1:20; sensitization, 60 minutes at 37°C.; ragweed pollen antigen, 4.5×10^{-3} μg./ml. *Exp. 030265.* Cells, PhNa; histamine available, 0.18 μg. per 1.1×10^7 wbc; serum, DiGu, 010865, 1:10; sensitization, 60 minutes at 37°C.; ragweed pollen antigen, 4.5×10^{-3} μg./ml.

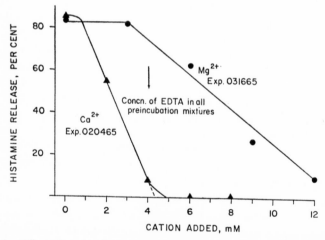

FIG. 4. Effect of preincubation with calcium or magnesium on the response of leukocytes sensitized *in vivo*. *Exp. 020465.* Cells, DeTy; histamine available, 0.24 μg. per 1.4×10^7 wbc; serum, DeTy, 020465, 1:20; incubation in serum, 60 minutes at 37°C.; ragweed pollen antigen, 2.5×10^{-5} μg./ml. *Exp. 031665.* Cells, TiPa; histamine available, 0.21 μg. per 1.2×10^7 wbc; serum, TiPa, 031665, 1:10; incubation in serum, 60 minutes at 37°C.; ragweed pollen antigen, 7.5×10^{-4} μg./ml.

doubt but that the potentiating effect of heparin on the sensitization of leukocytes is not associated with binding of calcium or magnesium. Thus, the action of heparin is particularly marked in the presence of EDTA, despite the lower concentration of heparin, ca. $10^{-5}\,M$, and its weaker ability to bind divalent cations (Levy and Osler, 1967a,b). For the data in Fig. 5, a histamine response of 40% was observed with a serum dilution of about 1:5 in the absence of additives. With EDTA this end point

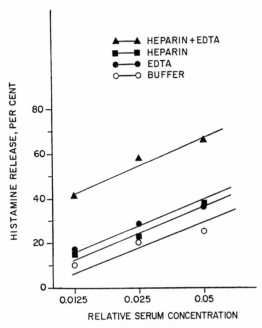

FIG. 5.   Passive sensitization of human leukocytes *in vitro*: Effect of heparin (10 μg./ml.), EDTA ($4 \times 10^{-3}\,M$), and a combination of the two on serum dose–response curves. *Exp. 121465.* Cells, JoGa; histamine available, 0.11 μg. per 7 $\times$ 10^6 wbc; serum, DiGu, 111665; sensitization, 120 minutes at 37°C.; ragweed pollen antigen, 1×10^{-2} μg./ml.

was more than doubled, to 1:12. With heparin alone in the sensitization mixture, a 1:14 serum dilution sufficed, whereas with both heparin and EDTA, the serum activity was further increased so that 1:80 dilution produced a 40% histamine response at optimal antigen levels. The values obtained in this experiment are not absolute, since the increment observed with heparin and EDTA varies substantially with cells from different donors.

The optimal quantity of heparin for potentiating leukocyte sensitization is about 10 μg./ml. of the serum–cell reaction mixture, but may range between 3 and 30 μg. The specific level seems to vary with the source of the leukocytes but the basis for this variation is not known. Other forms of heparin such as N-succinyl heparin and the synthetic compound, Mepesulfate, also promote sensitization but they are less effective, requiring higher levels, such as 100 μg./ml. In the thought that a heparin–fibrinogen interaction at the cell surface may have influenced the sensitization reaction, serum specimens were absorbed with BaSO$_4$ at levels of 40 and 300 mg./ml. This treatment did not diminish the enhancing action of heparin. Experiments designed to locate the site of heparin action indicated that this compound, like EDTA, was effective only when present during the cell-sensitization step. Pretreatment of either the cells or serum with heparin did not alter the subsequent response of the cells to antigen. Heparin enhances the sensitization of leukocytes in an unknown manner that may be related to the specific charge or configuration of cell membrane sites. Although the mechanism is unknown, it is of interest that Cohn and Parks (1967) observed a potentiating effect of heparin on pinocyte formation.

3. Donor's Cells—a Major Variable in the Sensitization Process

As noted by Middleton (1960), there is a marked variation in the capacity of leukocytes from different donors to bind the immunoglobulins which mediate the antigenic release of histamine. In fact, only about 20% of the individuals we have tested provide cells suitable for these studies, i.e., capable of a 50% histamine release by optimal levels of antigen (Levy and Osler, 1966). Many donors furnish cells that never release more than about 25% of their histamine store, and these relatively weak responses are not increased when heparin is incorporated in sensitizing reaction mixtures.

C. Mechanism of Passive Sensitization

It is premature to propose a definitive hypothesis regarding the mechanism of antibody fixation on the basis of these studies. Some of the obvious limitations flow from the use of heterogeneous suspensions of blood leukocytes as the responsive cells and whole serum as an antibody source. It may be suggested that the fixation of reagins is an energetic, time-consuming process involving specific receptors on the cell membrane and a unique conformation of the appropriate polypeptide chains of the immunoglobulins.

The potentiating effect of the polyanions such as heparin and EDTA also suggests that positively charged groups on the sensitizing antibodies play an important role in the interaction with the hypothetical cell receptors. In view of the pronounced chelating action of EDTA, this compound may promote passive sensitization by removing or binding positively charged residues on the cell membrane, thus rendering these sites more receptive to antibody. Incubation in EDTA has also been found to improve the efficacy of the cellular release mechanism, provided it is removed prior to the addition of antigen. Heparin, on the other hand, probably acts at a different site on the cell membrane than does EDTA. As is well known, heparin chelates cations quite poorly relative to EDTA, yet passive sensitization is markedly improved when both EDTA and heparin are available during the interaction of the leukocytes with the reaginic serum. A distinctive property of heparin is its extremely high anionic charge density, and it is this characteristic which may be important in enhancing passive sensitization. Thus, heparin and EDTA may act as ligands thereby increasing the number of effective antibody molecules per cell or by virtue of a change in the ability of the cell to respond to antigen more efficiently. Questions of this type can be approached more directly with purified preparations of human immunoglobulins capable of tissue or cell fixation. The need for a specific conformational requirement is suggested by several studies, typified by those of Ishizaka et al. (1961) and of Leddy et al. (1962) who showed that irreversible rupture of disulfide bonds in rabbit and human γ-globulins reduced their capacity for sensitization of the guinea pig and human skin. These disulfide bonds are probably localized in the Fc fragment of rabbit antibody H chains as indicated by the findings of Ovary and Karush (1961). Current attempts by Ishizaka and Ishizaka (1966) and Ishizaka et al. (1966) to implicate a specific immunoglobulin as the sole carrier of reaginic activity may provide the necessary reagent for further studies (cf., however, Yagi et al., 1963; Colquhoun and Brocklehurst, 1965).

Identification of Leukocyte-Sensitizing Activity As Reaginic Antibody

Evidence has been accumulated to support the view that the leukocyte- and skin-sensitizing activities in reaginic sera may be mediated by the same immunoglobulins. Both are thermolabile, being destroyed by heating at 56°C. for 2 hours (Levy and Osler, 1966). Both the leukocyte- and skin-sensitizing activities in serum are inactivated by reduction and alkylation (Leddy et al., 1962; Levy and Osler, 1967c). Comparative assays of skin- and leukocyte-sensitizing activities in fifteen sera yielded a rank–order correlation of 0.92, significant at $p < 0.001$.

Finally, leukocyte-sensitizing activities have been eluted from cross-linked dextran gel columns in a pattern similar to that for skin-sensitizing reagin, thus strengthening the likelihood that one class of immunoglobulins serves both the *in vivo* and the *in vitro* functions.

IV.   The Immune Release of Histamine from
Sensitized Human Leukocytes

A.   Nonimmunological Parameters

1.   Cell Types

Identification of the cell types involved in the histamine-release process and evaluation of their specific contributions to the total response has not yet been achieved. Erythrocytes, platelets, monocytes, and lymphocytes contain minimal quantities of histamine (Graham *et al.*, 1955). These investigators concluded that 95% of the human blood histamine is partitioned among the basophiles, eosinophiles, and neutrophiles at ratios of about 50:30:15. We have often observed that the immune release of histamine attains the same level that can be extracted from the cells by chemical means, i.e., 100%. Since the basophiles supposedly contain only half of the blood histamine, it may be assumed that the other granulocytes, e.g., eosinophiles and neutrophiles also participate in the allergic reaction. Direct evidence in this regard is still lacking. However, the conclusion is warranted that histamine does exist in a preformed state within human leukocytes. The heterogeneity of these cell populations poses a serious obstacle in these studies, and efforts to obtain more homogeneous cell suspensions are under way. Sorely needed in this regard is a reproducible procedure for the isolation of blood basophiles, in a physiologically competent state and free of the other granulocytic leukocytes.

2.   pH and Ionic Strength

The release of histamine from the cells of ragweed allergic donors is quite sensitive to changes in pH and ionic strength. Optimal responses are obtained within the pH limits of 7.3 to 7.4, and ionic strength in the range of $\mu = 0.14$ to 0.16. Changes of these experimental conditions in either direction depress the allergic response of human leukocytes. Austen and Brocklehurst (1961) and Austen and Humphrey (1962) noted greater release of histamine from perfused chopped guinea pig lung when the molarity of the NaCl in the ambient fluid was reduced to about 0.143. Species differences or the use of tissue fragments rather than of cell suspensions may account for these divergencies. However, the findings of Humphrey *et al.* (1963) with rat mast cells coincide with those for human leukocytes.

3. Temperature

As originally demonstrated by Schild (1939) in experiments with guinea pig aorta, optimal release of histamine from human leukocytes is also observed at body temperatures (Fig. 6). In fact, the human cells fail to respond if the temperature is lowered to about 20°C. At 30°C. considerable release is obtained, but the latent period is much longer than at 36.4°C., suggesting a high-temperature coefficient for

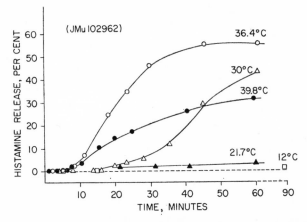

FIG. 6. Time course of histamine release at different temperatures. Ragweed pollen antigen E, 5×10^{-5} µg./ml.; 1.4×10^7 wbc; histamine available, 0.21 µg.

one or more reaction steps preceding histamine release. When cells are exposed to 39.8°C. for 20 minutes, the lag phase is still the same as that seen at 36.4°C., but the rate and extent of the reaction are depressed. Thus, the release phase is also temperature-sensitive, and the deleterious effects can be detected even at 40°C. (cf. also Mongar and Schild, 1957).

4. Divalent Cation Requirements

As illustrated in Fig. 7, the release of histamine by antigen is not observed until adequate levels of calcium and magnesium are incorporated into the reaction medium. In the absence of Mg^{2+} the response increases once Ca^{2+} levels exceed $1.0 \times 10^{-4} M$, until peak values are attained in the range of 6 to $10 \times 10^{-4} M$ of Ca^{2+}. However, the quantities of histamine released with Ca^{2+} as the sole divalent cation rarely exceed 60%. The response observed with Mg^{2+} in the absence of Ca^{2+} does not exceed 20%. When both cations are present, the immune release of histamine is greatly potentiated even at subthreshold levels of each metal. The requirement for both Ca^{2+} and Mg^{2+} in this allergic reaction is thus readily demonstrable (cf., however, Mongar and Schild, 1958, 1962; Chakravarty, 1960). Concentrations of calcium exceeding $10^{-3} M$ tend to depress the response.

The sequence of action of these metals on human leukocytes has not been defined. In experiments of the type summarized in Fig. 8, the

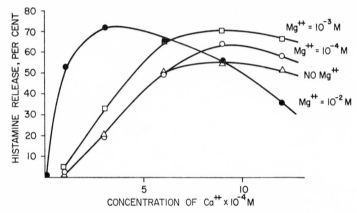

FIG. 7. Histamine release as a function of calcium and magnesium concentrations. Histamine available, 0.15 µg. per 1.3×10^7 wbc.

chelating agent EDTA was added about 22 minutes after the addition of antigen, and after about 65% of the available histamine had been released. The addition of EDTA promptly terminated any further translocation of histamine. The process could be resumed in the presence of a slight excess of Ca^{2+}, but not of Mg^{2+}. When Ca^{2+} is added in excess of the EDTA concentration, the reaction medium will contain the added cation plus a small quantity of Mg^{2+} which is released from its binding to EDTA by Ca^{2+}. It cannot, therefore, be concluded that the resurgence of the histamine response which follows the addition of Ca^{2+}, as in Fig. 8, is due solely to this metal. Additional studies are underway in an attempt to define the sequence of divalent cation participation in the immune release of histamine.

5. The Effect of Serum Albumin

The decision to incorporate 0.03% of HSA in the buffered diluent was based on the following considerations. The inclusion of an indifferent protein in the cell–antigen–buffer reaction mixture was considered advisable to protect the cells and the minute quantities of antigen (10^{-7}–10^{-1} µg./ml.) required to initiate histamine release from denaturation at air–liquid and liquid–glass interfaces. Numerous titrations showed that the amounts of histamine in the fluid phase of reaction mixtures diminished as the concentrations of HSA were raised. The mechanism of this effect remains to be explored, but the concentration adopted for the present investigation, 0.03%, does not depress the response and is sufficiently high to prevent unpredictable antigen losses due to denaturation.

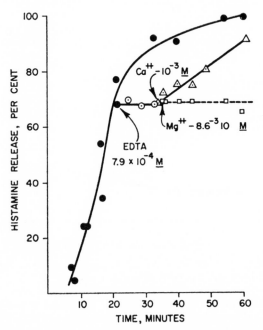

Fig. 8.   Inhibition of histamine release from human leukocytes by EDTA. Reversal by calcium but not by magnesium. Ca^{2+} in buffer, 6×10^{-4} M. *Exp. 100763.* Cells, BeWi; 0.16 µg. histamine per 10^7 wbc.

A finding of some interest, and possibly related to the action of serum albumin, is described in Fig. 9. Depicted there are the histamine release values obtained from a single cell suspension at two antigen levels (1.67 and 5.0×10^{-4} µg./ml.), after 2 and 5 cycles of centrifugation and resuspension in tris buffer containing Ca^{2+}, Mg^{2+}, and 0.03% HSA. The term "cell extract" in this figure refers to the supernatant of reaction mixtures (containing tris buffer with 1.0 mg. albumin per milliliter) which had been incubated for an hour at room temperature with human leukocytes isolated from either ragweed-sensitive or normal donors in the customary manner.

As seen in Fig. 9, repeated cell washing lowers the responsiveness of the cells without depleting their store of histamine. Incorporation of the cell extract restores the histamine-releasing potential. Current information regarding the nature of the cell extract is limited to the following rudimentary observations. The enhancing activity may be obtained from the cells of allergic or nonallergic individuals. It is not dialyzable and resists boiling in water for 1 hour. Characteriza-

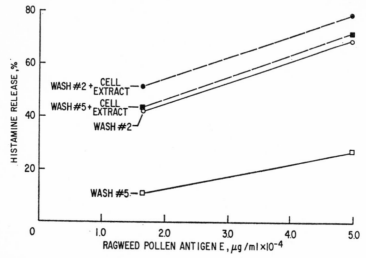

FIG. 9. Reduction in histamine release following repetitive cycles of washing and centrifuging. Restoration by leukocyte cell extract.

tion of this material may prove helpful in advancing present understanding of the mechanism of immune histamine release from human leukocytes.

B. ANTIGEN REQUIREMENTS

1. In the Absence of Normal Serum

Representative antigen dose–response relationships with cells from three ragweed-sensitive donors are given in Fig. 10. The results observed with the cells of LML and CMcA are the more typical. In studies with several hundred different cell populations, threshold antigen levels have been found to range from 10^{-7} to 10^{-3} μg./ml. Peak histamine responses of 70 to 100% are obtained with about 10 to 30-fold greater quantities.

As in other allergic reactions, excessive amounts of antigen depress the response (VanArsdel *et al.*, 1958; Osler *et al.*, 1959a; Frick *et al.*, 1965). The findings observed with cells DD (Fig. 10) typify about 20% of donors whose leukocytes, following interaction with antigen, yield only a small fraction of the histamine that can be obtained from acid lysis of these cells. The nature of this limitation is obscure.

The data in Fig. 11 describe selected portions of dose–response curves obtained with cells from ten ragweed-sensitive individuals. A linear

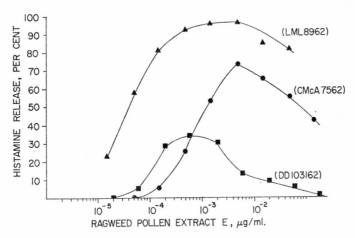

FIG. 10. Histamine release from ragweed-sensitive leukocytes obtained from three donors. LML—1.8×10^7 wbc; histamine available, 0.38 μg. C McA—2.2×10^7 wbc; histamine available, 0.42 μg. DD—2.3×10^7 wbc; histamine available, 0.21 μg.

relationship characterizes the response in the range of 20 to 80% histamine release. These linear segments have similar slopes and are displaced laterally with respect to the concentration of ragweed antigen. These data indicate that a quantitative expression of cell sensitivity can be employed for comparative purposes, in terms of the quantity of ragweed antigen required to elicit a 50% response. As will be developed later, this expression of cell sensitivity (G_{50}) is inversely related to the severity of symptoms experienced by the cell donor during the ragweed pollen season.

The *in vitro* reaction system utilizing leukocytes from allergic donors is responsive to minute quantities of ragweed pollen antigen. Considering the data in Fig. 11 and the estimated molecular weight of this antigen at about 38,000 (King and Norman, 1962), it may be calculated that 50% histamine release requires the presence of about 10^6 antigen molecules for cells of the most sensitive donors. Since the average cell population contains about 0.2 μg. histamine per 10^7 polymorphonuclear leukocytes, each molecule of the added antigen accounts for the release of 10^6 or more molecules of histamine. This number may be too low by at least one order of magnitude. We have been able to recover virtually all of the antigenic input from supernates of antigen–cell reaction mixtures when these were maintained at 15°C. to permit antigen–antibody interaction in the absence of histamine release.

The data in Fig. 12 describe an experiment of this type. It is apparent that the dose–response curve obtained with the supernatant fluids of

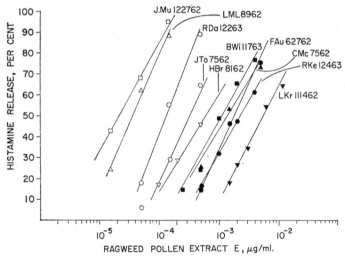

FIG. 11. Dose–response relationships for histamine release as a function of antigen concentration.

prior reaction mixtures carried out at 15°C., is virtually superimposable on that obtained with solutions of antigen not previously admixed with cells from allergic donors. On these grounds, the amount of antigen in firm union with the cell-bound antibodies does not seem to exceed the experimental error of the method, i.e., 10–20% in terms of antigen concentration (Lichtenstein and Osler, 1966b). Thus, minimally effective antigen levels for histamine release from 10^7 leukocytes of the most sensitive donors approximate 10^{-7} μg. or 10^{-17} mole.

2. Effect of Normal Human Serum

The discussion thus far has been concerned with the interaction of ragweed antigen and responsive leukocytes, in a medium devoid of serum proteins except for those present on the cell surfaces and the 0.03% HSA additive. It was of interest from several viewpoints to define the possible role of human serum from nonallergic individuals (Lichtenstein and Osler, 1966a).

The data in Fig. 13 show that normal human serum can potentiate the release of histamine, provided that the extent of the reaction is otherwise limited by suboptimal antigen levels. When the amounts of antigen are adjusted for a maximal cell response, both the rate and

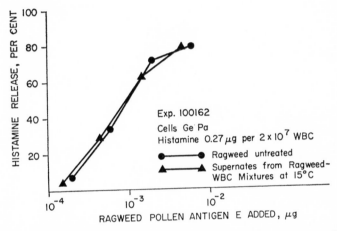

FIG. 12. Demonstration that free antigen remains in the supernate of reaction mixtures containing ragweed antigen and sensitive leukocytes. Each point on the curve (▲) is plotted in terms of the amount of histamine released by the level of antigen in the supernate, assuming no antigen binding by the cell. These results are compared with a dose–response curve utilizing antigen not previously in contact with leukocytes (●).

extent of histamine release are unaffected by the presence of serum in the fluid phase. With lower levels of antigen, each of these parameters is increased.

The potentiating effect of normal serum inevitably suggested that C′ may participate in this reaction. The results of pertinent experiments in this regard are clear (Lichtenstein and Osler, 1966a). The heightened responses observed in the presence of normal serum cannot be attributed to the contribution of the serum C′ system for the following reasons (cf. also Middleton and Sherman, 1960).

Serum heated to 56°C. for 60 minutes, even after subsequent treatment with NH_4OH, is generally more effective than an unheated aliquot of the same specimen. The C′1a and C′4 activity levels of heated and NH₄OH-treated serum samples can be decreased to 0.03 and 0.2% of those found in the untreated serum, yet the potentiating effect of the former may exceed that of the fresh serum (Lichtenstein and Osler, 1966a). Since the enhancement by untreated serum is not usually demonstrable below a concentration of 1%, it does not seem profitable to pos-

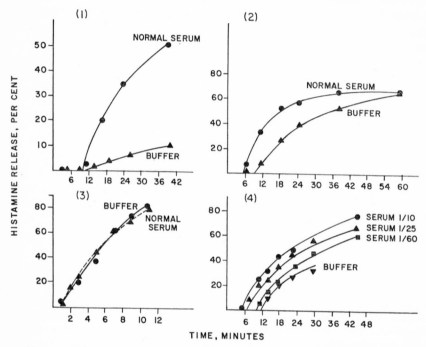

Fig. 13. The variable effects of normal human serum on the release of histamine from leukocytes by ragweed pollen antigen E. 1—At highly limiting antigen levels (Exp. 060463) F GI; 2—at moderate antigen levels (Exp. 062563) I Si; 3—at optimal antigen levels (Exp. 080963) R Da; and 4—at varying serum dilutions (Exp. 082664) T Pa.

tulate that the residual C′ activity in the heated and NH₄OH-treated serum accounts for the augmented response, unless the leukocytes themselves contribute the missing components. That this is probably not the case is demonstrated by the data in Fig. 14. For this experiment, leukocytes were washed in tris buffer containing 0.05 M EDTA, thereby reducing the C′1a activity levels by more than 99%. The response of these cells could be augmented by fresh or heated normal human serum, the latter being more effective. Furthermore, leukocytes washed in EDTA were at least as responsive as other cell aliquots washed concurrently in tris buffer devoid of EDTA, and, therefore, carrying considerably more C'1a activity.

The essence of these experiments is to render unlikely the possibility that C' components participate in the potentiation of histamine release by normal serum when antigen levels are limiting. However, the data cannot be interpreted as excluding the participation of components of the C' system in the leukocyte histamine release process. Some of the evidence along these lines is summarized below.

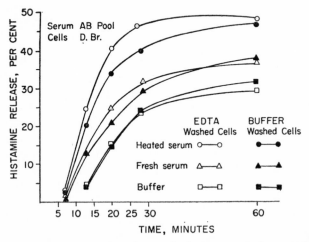

Fig. 14. Histamine release from human leukocytes washed in tris buffer or in EDTA–tris buffer. Enhancement by fresh or heated normal serum. *Exp. 051764.* Cells, D. Br; 0.25 μg. histamine per 1.3×10^7 cells.

It is pertinent to note that marked differences have been reported with respect to the serum requirements of other *in vitro* reaction systems used for the study of immediate-type allergic responses. Thus, the addition of serum is not essential for the release of vasoactive compounds from guinea pig smooth muscle (Dale, 1913; Neu *et al.*, 1961; Binaghi *et al.*, 1962) or minced guinea pig lung preparations (Mongar and Schild, 1956; Brocklehurst *et al.*, 1961). It is, however, difficult to exclude an endogenous serum contribution when intact tissues serve as the reservoirs of vasoactive compounds. The requirement for serum factors in the release of vasoactive compounds from rat mast cells has also been somewhat controversial (Archer, 1960; Keller, 1966; Austen and Humphrey, 1963; Humphrey *et al.*, 1963). Most recently, Becker and Austen (1966) and Austen and Becker (1966) have distinguished between non-C'-dependent histamine release from sensitized rat peritoneal mast cells by specific antigen and C'-dependent histamine release from normal rat peritoneal mast cells by rabbit antibody to rat γ-globulin. In contrast to the observations with rat peritoneal mast cells, Secchi *et al.* (1967) have found that human leukocytes release significant amounts of histamine in the reaction with rabbit anti-human γ-globulin in the absence of added serum components, much in the same manner as do leukocytes from ragweed-sensitive donors in the presence of specific antigen (Osler *et al.*, 1964). Since histamine is probably not the sole mediator of immediate allergic phenomena in rats (Brocklehurst *et al.*, 1955; Stechschulte *et al.*, 1967), the serum requirements for the release of other vasoactive compounds require additional study.

Several years ago evidence was offered to suggest that serum C' components may govern the intensity of wheal and flare reaction induced in the rat by heterologous antibody (Bier *et al.*, 1955; Osler *et al.*, 1957). These findings have now been confirmed by Stechschulte *et al.* (1967). However, passive cutaneous anaphylactic reactions in the guinea pig are mediated by homologous immunoglobulins which do not interact with the C' system (Ovary *et al.*, 1963; Bloch *et al.*, 1963). Finally, serum factors are essential for histamine release from rabbit platelets (Humphrey and Jaques, 1955; Barbaro, 1961; Gocke and Osler, 1965; Siraganian *et al.*, 1967).

One further comment may be pertinent in this context. Recent developments in C' technology have demonstrated the presence of reaction products formed as a consequence of C'-component activity (Mayer, 1965; Stroud *et al.*, 1966; Müller-Eberhard, 1966). Some of these may possess biological activity. The recognition that anaphylatoxin (Friedberger, 1909; reviewed in Osler *et al.*, 1959b; Dias da Silva and Lepow, 1965; Jensen, 1967) and chemotactic factors (Ward *et al.*, 1965) are products of C'-component interaction, are timely examples. Still others may be found as this phase of C' studies progresses, and the study of their possible implication in cellular mechanisms of histamine release may prove interesting. The growing recognition that different immunoglobulins mediate the release of various vasoactive compounds also provides an ample basis for additional studies. The brief discussion of this question, which has not touched upon problems of species and target-cell variations, clearly indicates the danger of generalizing about C' participation in allergic responses.

C. THE FATE OF HUMAN LEUKOCYTES IN THE
    IMMUNE RELEASE OF HISTAMINE

The release of histamine or serotonin from tissue mast cells is usually associated with degranulation phenomena and possible loss of cell viability (Fawcett, 1954). Indeed, the immune mediator of degranulation was originally called the "mast cell lytic antibody" (Mota, 1963). Our observations with human leukocytes coincide with the suggestions of Smith (1958a,b). He reported that the release of histamine from rat peritoneal mast cells, due to protamine sulfate and toluidine blue, was not necessarily associated with cell disruption and death (Smith, 1958a,b, 1963). The reports by Audia and Noah (1961) and Noah and Brand (1963) also indicate that the immune release of histamine need not involve irreparable damage to the cell engaged in this process. We have observed that human leukocytes can release virtually all of their histamine without signs of cell damage as judged by motility, uptake of vital dyes, or disrupted membrane function. On this basis we have been led to believe that the histamine-release process is essentially an exacerbation of a normal secretory function. There are at least two reasons why our findings differ from the numerous reports which equate the release of pharmacologically active compounds from rat mast cells with degranulation phenomena. The first concerns the differences attributable to the use of human peripheral blood leukocytes rather than rat mesentery mast cells. Second, the quantities of antigen required to mediate the release of histamine from human leukocytes, i.e., 10^{-7}–10^{-3} μg., are generally a million times lower than those used in the rat mast cell studies. In almost all of the studies with rat cells, antigen levels in the milligram range have been employed.

Of interest in this context are the data in Fig. 15 which show that the addition of EDTA at any stage in the immune release process promptly terminates the reaction. It appears then that the translocation of histamine across the granule and cell membranes requires divalent cation interaction. These findings are not consistent with the concept of irreparable cell membrane damage which characterizes the terminal phases of immune cytotoxic phenomena as mediated by specific antibody and C' (Green and Goldberg, 1960). In fact, the available data suggest that only a physiologically competent leukocyte can participate in the immune release of histamine. As mentioned above, exposure to a temperature of 40°C. for 20 minutes diminishes the cell's ability to respond, and complete loss of allergic reactivity occurs at 45°C., as previously

noted by Schild (1939). Further indication of the need for viable leukocytes was obtained in experiments with the esterase inhibitor, diisopropyl-
fluorophosphate (DFP). Incorporation of this cell poison into ragweed
pollen–leukocyte reaction mixtures at levels approximating $5 \times 10^{-4} M$
inhibited 50% of the response observed with control cell suspensions.
Similar degrees of inhibition were obtained when the leukocytes were
preincubated with 2×10^{-4} mole of DFP and subsequently washed. The
diminished response could not be restored with a broad range of
antigen levels. Salicylaldoxime, an inhibitor of immune hemolysis (Mills
and Levine, 1959), yielded similar results. It was concluded that the

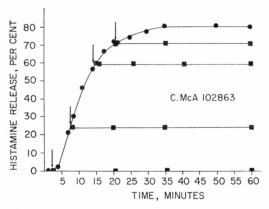

Fig. 15. Effect of EDTA addition on the release of histamine from human
leukocytes by ragweed pollen antigen E. EDTA (0.02 M final concentration) added
at times indicated by arrows.

interference with histamine release by these two compounds need not be
due to inhibition of an antigen-activated reaction step but may involve
an alteration of cellular homeostasis (Lichtenstein and Osler, 1966b, cf.
also Becker and Austen, 1966). This interpretation coincides with findings obtained with guinea pig tissues by Mongar and Schild (1956).

The suggestion that the immune release of histamine requires a
metabolically active leukocyte was further strengthened by the results
of experiments dealing with the efflux of ^{42}K. Suspensions of leukocytes
from ragweed-sensitive donors were equilibrated with this isotope, then
washed, and reacted with antigen. As seen in Fig. 16, prompt and
marked loss of intracellular histamine occurs without a concomitant and
enhanced efflux of ^{42}K (Lichtenstein and Osler, 1966c). These results
may again be contrasted with those recorded for C'-mediated immune
cytotoxic phenomena, wherein early and rapid loss of intracellular ^{42}K

was observed (Green and Goldberg, 1960). This observation has been extended in more detailed experiments involving the interaction of rabbit antiserum to human γ-globulin and leukocytes from normal or allergic donors (Secchi *et al.*, 1966). Under these conditions also, histamine may be released without evidence of increased K^+ efflux.

There is one consideration that must be carefully evaluated before final acceptance of the interpretations drawn from the experiments comparing histamine release with K^+ efflux. The reaction system under

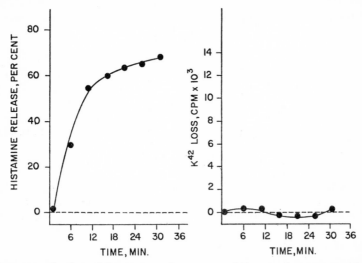

Fig. 16. The loss of histamine and potassium following the exposure of sensitive human leukocytes to ragweed antigen E (0.001 μg./ml.). *Exp. 011365*. Cells of J B1 in 1:10 normal serum.

study comprises a heterogeneous population of cells differing in their content of histamine and, possibly, in the number and types of immunoglobulins which they carry. As already noted, Graham and her co-workers (1955) have reported that about half of the cellular histamine in human blood is confined to the basophiles which constitute less than 1% of the blood leukocytes. If it is assumed that all the leukocytes contain equivalent amounts of K^+, an allergic event involving only the basophiles would not result in a discernible and increased ^{42}K efflux, despite the demonstrable histamine response. This limitation loses some of its cogency since we have thus far failed to observe an enhanced efflux of ^{42}K from cells that released more than 90% of their histamine store.

D. FURTHER CONSIDERATION OF THE REACTION MECHANISMS
INVOLVED IN HISTAMINE RELEASE

On a rational basis, the immunologically mediated release of histamine would seem to require several reaction steps. There is, of course, the triggering immune event which probably occurs at the cell surface. This event activates at least two other reaction steps involving the movement of the histamine-laden cytoplasmic granules to the cell surface and the cell membrane alterations which lead to the release of histamine. The observations regarding the need for metabolically active cells sup-

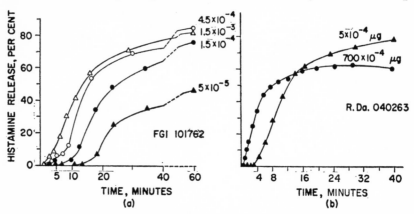

FIG. 17. (a) Effect of antigen concentration on time course of histamine release by ragweed pollen antigen E. Histamine available, 0.27 μg. per 1.2 × 10⁷ wbc. (b) Time course of histamine release at optimal and antigen excess (inhibitory) concentrations. Lag period: 5 × 10⁻⁴ μg. ragweed pollen antigen E per milliliter, 3.75 minutes; 700 × 10⁻⁴ μg. ragweed pollen antigen E per milliliter, 0.50 minutes. Histamine available, 0.30 μg. per 1.6 × 10⁷ wbc.

port this type of reaction mechanism (Copenhaver *et al.*, 1953; Mongar and Schild, 1956; Lichtenstein and Osler, 1964). The multistep nature of the release process can also be deduced from the following data.

1. Lag Phase

The curves plotted in Fig. 17 show that a lag phase intervenes between the addition of ragweed pollen antigen to a suspension of responsive cells and the detection of histamine in the fluid phase. This interval may include the time necessary for the formation of antigen–antibody complexes of appropriate configuration at the cell surface. Inhibition of the response by excess antigen is in accord with this notion. The experiments described in Figs. 18 and 19 indicate the need to consider additional factors. The solid circles in curve A of Fig. 18 outline the course of histamine release for a cell population with an antigen level of 0.01 μg./ml. The solid triangles in curve C show the marked suppression in rate and extent of this reaction when cells and antigen are admixed at 37°C., in the absence of calcium and magnesium. Without these cations the immune event can presumably run its course but fail to initiate the subsequent steps required for histamine release. Further evidence for

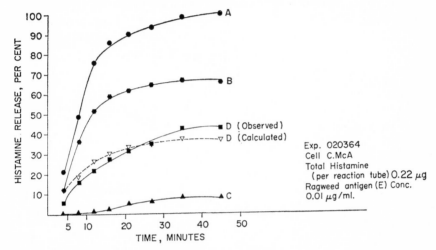

FIG. 18. Histamine release from human leukocytes. Curve A—cells preincubated in buffer for 40 minutes at 20°C. before addition of ragweed to a final concentration of 0.01 μg./ml. and incubation at 37°C. B—Cells preincubated with ragweed (0.02 μg./ml.) for 40 minutes at 20°C. before adjustment of ragweed concentration to 0.01 μg./ml. and incubation at 37°C. C—Cells preincubated with ragweed (0.02 μg./ml.) for 40 minutes at 37°C. without added Ca^{2+} and Mg^{2+}, before addition of these cations at zero time, adjustment of ragweed concentration to 0.01 μg./ml., and subsequent incubation at 37°C. D (observed)—Equal volumes of cells from flasks A and C were mixed, the ragweed concentration adjusted to 0.01 μg./ml. and then incubated at 37°C; D (calculated)—arithmetic means of data from curves B and C.

the multistep nature of this reaction is provided by the data summarized in curve B which demonstrate the temperature dependency of the reaction step(s) set into motion by the antigen–antibody reaction. Thus, each of the cell populations (B and C) had been affected by the experimental pretreatment, either at the temperature- or cation-dependent step of the sequence. It is possible that the temperature-dependent step coincides with one of the cation-requiring reactions, but this remains to be established (see curve D, Fig. 18).

Recently, Pruzansky and Patterson (1967) have reported the isolation of an intermediate stage in the immune release of histamine from human leukocytes. In their hands, cells treated with antigen and washed could be restored to full histamine-releasing potency upon the restoration of optimal levels of calcium and magnesium. Additional studies of this type and those summarized in Fig. 18 may clarify the nature of the intermediates involved in histamine release (see curve D, Fig. 18).

The data in Fig. 19 show that pretreatment of the leukocytes with subthreshold amounts of ragweed diminishes the histamine-releasing potential of the cell, even when optimal experimental conditions are restored. The mechanistic possibilities raised by these experiments can

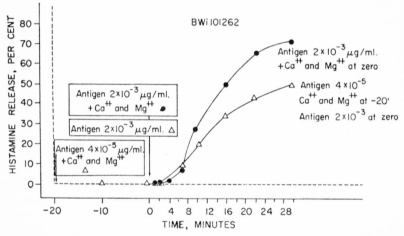

FIG. 19. Effect of preincubation of ragweed-sensitive cells with subthreshold quantities of antigen E in the presence of divalent cations at 37°C. Histamine available, 0.22 μg. per 1.2×10^7 wbc.

only be speculated upon at this time. A likely possibility concerns the structural property of the immune complex which mediates histamine release. There is ample evidence to indicate that only immune complexes of specified configurations can initiate the release process. The antigen–dose response curves suggest that complexes formed with subthreshold or excessive amounts of antigen are less efficient than those formed with optimal antigen levels. The nature of the complex which is most efficient on a cell surface need not conform to the stable three-dimensional complexes achieved in the equilibrium state of antigen–antibody interaction. The latter reflect fluid phase events whereas histamine release is initiated at a cell surface.

2. Leukocyte-Borne Immunoglobulins

The immunoglobulins that mediate the antigenic release of histamine from cells of allergic donors have not been identified in these studies, but there is reason to believe that they are probably identical with the skin-sensitizing reagins (cf. Levy and Osler, 1967c). It is abundantly clear that human leukocytes bear antibodies of varying specificities, as demonstrated by the ability of many different antigens to release histamine. Noah and Brand (1963) have described positive reactions with a variety of different allergens, an observation confirmed by Pruzansky and Patterson (1966) and by May et al. (1967). We have also detected reactive immunoglobulins on leukocytes to the following

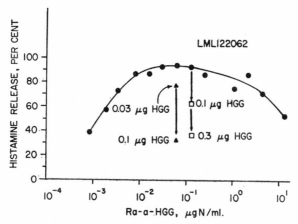

FIG. 20. Histamine release from human wbc by rabbit antihuman γ-globulin; inhibition by human γ-globulin.

antigens: Aqueous extracts of schistosomal antigens (adult and cercarial forms), *Staphylococcus* and tetanus toxoids, *Salmonella* lipopolysaccharides, and high molecular weight dextrans.

Assays of C' fixation with rabbit antiserum to human γ-globulin indicated that human leukocytes carry γ-globulin even after 6 cycles of centrifugation and resuspension. On the basis of preliminary estimates, it appears that 10^6 peripheral blood leukocytes carry the equivalent of 0.01 μg. of human γ-globulin N or an average of about 10^5 molecules per cell. The granulocytes in these cell suspensions release histamine in the presence of specific rabbit antiserum to human γ-globulin, as summarized in Fig. 20 (Osler et al., 1964; Secchi et al., 1966). It will be of considerable interest to repeat these studies with more homogeneous cell suspensions. Although the structural relationships of these cell-bound immunoglobulins to the cytophilic antibodies, described by Boyden and Sorkin (1960), Uhr (1965), and Berken and Benacerraf (1966) remain to be explored, the functional role of the leukocyte-bound immunoglobulins in the histamine release reaction seems to have been unequivocally established.

V. Some Applications of the *in Vitro* Studies to the Biology of Ragweed Pollinosis

The procedures described above have been applied to immunological and clinical studies of ragweed hay fever. As will be developed below, it appears that the *in vitro* model can be utilized to describe the important immunological parameters of ragweed pollinosis. The basic reason for the validity of the *in vitro* model stems from the fact that the responsive leukocyte serves as the carrier of the etiologic agent—the reaginic antibody. The interaction of these cells with the allergen has been utilized in several ways: to assay blocking and reaginic antibody activities, to obtain an index of clinical disease severity, to trace the course of reagin formation, and to assay the allergenic capacities of different antigens.

A. Assay Procedures

1. Assay of "Blocking" Antibodies in Terms of Antigen-Neutralizing Capacity

The early studies of Cooke and his colleagues (1935, 1937) and of Loveless (1940) demonstrated that the sera of hay fever patients manifest at least two biological activities—the blocking antibodies which follow parenteral antigen administration and the naturally occurring skin-sensitizing reagins. These findings have been extensively confirmed in human skin test experiments and have now been extended to the *in vitro* reaction system involving leukocytes from ragweed-sensitive donors (VanArsdel and Middleton, 1961; Matthews *et al.,* 1962; Lichtenstein and Osler, 1964, 1966a,b; Levy and Osler, 1966, 1967b). This extension has included the development of a procedure to estimate the antigen-neutralizing capacity (ANC) in human allergic serum. Preincubation of ragweed pollen with the serum from a ragweed-sensitive donor diminishes the histamine-releasing activity of the antigen. Since the extent of histamine release can be quantitatively related to the amount of antigen (Lichtenstein and Osler, 1964), serum assays for ANC become practicable.

These assays are performed by incubating several levels of antigen with serum from an allergic and a normal donor. Reactive leukocytes from allergic donors are introduced upon completion of the fluid phase events and these cells provide the means of comparing the histamine-releasing activity of the antigen after interaction with the allergic donor serum and with the normal serum. In view of the potentiating effect of normal serum, all assays are carried out in a buffer diluent containing serum from a nonallergic individual at a concentration of 10%. Representative dose–response curves which describe the release of histamine from reactive leukocytes in the presence of normal and of allergic human serum are depicted in Fig. 21. The effect attributable to antibody in the

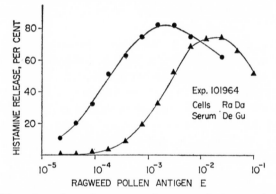

FIG. 21. Histamine release from ragweed-sensitive leukocytes by ragweed antigen E in the presence of normal human (●) and allergic human serum (▲), both at a final dilution of 1:10.

allergic serum is shown by a lateral shift of the curve toward higher antigen levels and by an increase in the threshold level of antigen needed for a perceptible response. The two curves parallel each other in the range of partial response (20–70%). For the data in Fig. 21, 50% of the cellular histamine store can be released by 1.7×10^{-4} μg. of antigen in the presence of normal serum. With the allergic serum, also at a 1:10 dilution, a 16-fold increase in antigen was required to achieve the same response. Suppression of histamine release by excess antigen is clearly noted in both curves. The parallelism of the two curves provides a key feature regarding the mode of action of allergic serum since it signifies that a fixed percentage, rather than a constant amount of antigen is bound. In Fig. 21 slightly more than 93% of the antigen input was rendered ineffective by the 10% dilution of this allergic serum. The percentage of antigen neutralized remained at 93%, even when the quantity of antigen rendered inactive ranged from 8.4×10^{-4} μg./ml., at 20% release, to 57.7×10^{-4} μg./ml. at a 70% response.

These findings recall the "percentage law" which has been invoked to describe the neutralization of virus particles by antibody when the latter is in marked excess (Andrewes and Elford, 1933). As currently interpreted, the outcome of the virus–antibody reaction is but a special case of the mass action law, under conditions in which a sufficient excess of antibody assures that no appreciable decrease of free antibody will occur as a result of the interaction with antigen (de St. Groth, 1962). The suggested analogy of the virus–antibody reaction to the ragweed pollen immune system is further demonstrated by the data in Table II, obtained after equilibrium conditions prevailed with respect to the fluid phase antigen–antibody reaction (Lichtenstein and Osler, 1966b).

For these experiments two sera from ragweed-sensitive donors were incubated with appropriate quantities of antigen. The mixtures were then assayed for residual activity with leukocyte suspensions from two ragweed-sensitive donors. The cell populations selected differ substantially in their G_{50} normal human serum values, i.e., the quantity of antigen required for 50% histamine release in the presence of normal human serum (cf. cells LML and C McA in experiment 082164, Table II). As noted previously, the action of allergic serum is reflected in the increased antigen requirement for 50% histamine release, e.g., the immune serum ViMa increased the G_{50} value of cells LML from 1.5 to 2.8×10^{-5} μg./ml. With cells from C McA, which are far less reactive than those of LML (72 and 1.5×10^{-5} μg./ml., respectively), the ViMa serum increased the antigen requirement from 72 to 140×10^{-5} μg./ml. The effect of serum ViMa was quite the same whether cells LML or C McA served as indices for the neutralization reaction, yielding 46 and 49% neutralization, respectively. These results, and the others in columns 7 and 8 of Table II, indicate that the percentage of antigen which is neutralized by an allergic human serum, remains fairly uniform despite great differences (10,000-fold for the data in Table II) in the sensitivity of the indicator cells used to estimate the activity of the residual, free pollen antigen. The observation that the percentage of antigen bound is unrelated to the quantity of antigen input, is incompatible with any quantitative relationship between antigen and antibody in the fluid phase except that in which the antibody is present in fairly large excess.

From a practical viewpoint, these data illustrate the basis for the *in vitro* assay of the ANC in serum specimens of ragweed-sensitive subjects. The rationale for this assay may be discussed in terms of the

TABLE II

The Percentage of Antigen Bound by Allergic Human Serum As Determined in Experiments with Leukocytes of Varying Reactivity to Ragweed Antigen

(1)[a]	(2)	(3)	(4)	(5)	(6)	(7)	(8)
Experiment	Leukocyte donor	G_{50} (NHS[b]) × 10^{-5} μg./ml.	Allergic serum donor	G_{50} (AHS[c]) × 10^{-5} μg./ml.	Quantity of antigen neutralized × 10^{-5} μg. (5)−(3)	% Antigen neutralized (6)÷(5) × 100	% Antigen-free
081364	G Pa	8.3	R Da	69	60.7	88.0	12.0
	LML	1.7	R Da	16	14.3	89.4	10.6
	G Pa	8.3	C McA	360	352	97.8	2.2
	LML	1.7	C McA	43	41.3	96.0	4.0
082164	LML	1.5	ViMa	2.8	1.3	46.4	53.6
	C McA	72	ViMa	140	68	48.6	51.4
	LML	1.5	R Ka	23	21.5	93.5	6.5
	C McA	72	R Ka	550	478	86.9	13.1
102164	LML	1.0	F Gl	6.8	5.8	85.3	14.7
	DST	150	F Gl	2800	2650	94.6	5.4
	LML	1.0	G Pa	25	24	96.0	4.0
	DST	150	G Pa	6000	5850	97.5	2.5
110464	J Mu	0.8	J Mu	9.7	8.9	91.8	8.2
	DST	180	J Mu	3700	3520	95.1	4.9

[a] Antigen and serum incubated together for 60 minutes at 37°C. prior to the addition of this mixture to the leukocyte suspension.
[b] Antigen required for 50% histamine release in the presence of normal human serum (1:10).
[c] Antigen required for 50% histamine release in the presence of allergic human serum (1:10).

data supplied by the experiment in Fig. 22. Two serum specimens were obtained from patient JBA, before and after a series of therapeutic injections of an aqueous extract of ragweed. The neutralizing capacity of each serum was studied at dilutions 1:40, 1:20, and 1:10. These were preincubated with several quantities of antigen and then the serum–antigen mixtures were added to ragweed-sensitive leukocytes for the histamine-release reaction. The response of these cells to normal human serum was also evaluated. As seen in Fig. 22, 2×10^{-5} μg. of antigen

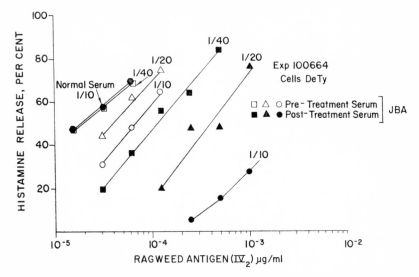

Fig. 22. Comparison of antigen-neutralizing capacities in three dilutions of pre- and posttreatment allergic human sera. The final serum concentration in each case has been adjusted to 10% with normal human serum.

yielded a 50% response in normal human serum as well as in a 1:40 dilution of the preimmunization allergic human serum. When the concentration of the latter was increased to 1:20 and to 1:10, the corresponding antigen levels for 50% release were 4 and 7×10^{-5} μg./ml. Following specific desensitization, the patient's serum neutralized much more antigen, the values for 50% histamine release being 1, 3.6, and 36×10^{-4} μg./ml. for the 1:40, 1:20, and 1:10 serum dilutions. The neutralizing index of allergic human serum (AHS) is expressed as the ratio G_{50} (AHS/NHS), i.e., the ratio of antigen required for 50% release in the test serum and normal human serum (NHS) as illustrated in Figs. 23 and 24. An important limitation of this assay emerges when the data comprising Fig. 22 are plotted as in Fig. 24. It is readily apparent

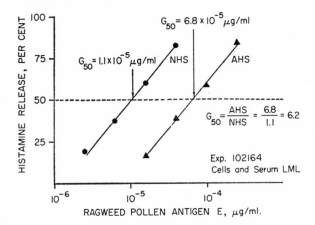

FIG. 23. Dose–response curves of one individual's cells in 10% normal (NHS) and autologous allergic (AHS) human serum. The method of calculating the level of antibody activity, or the G_{50} (AHS/NHS), is indicated.

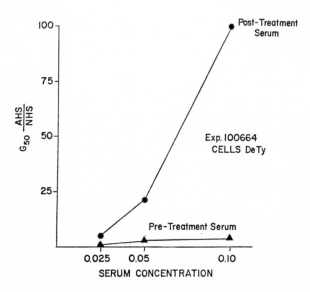

FIG. 24. The quantity of antigen required for 50% release of histamine in the presence of allergic human serum as compared to that for normal human serum. The ratio of these two values is designated G_{50} (AHS/NHS) and is plotted for each of three serum dilutions.

that the relative potencies of the pre- and posttreatment sera are 5, 9, or about 30, depending upon which dilution of the two sera is selected for comparison. The discriminating ability of the method increases at the higher serum concentrations. Generally, serum levels of 10% are compared since the use of higher concentrations is not practicable (Lichtenstein and Osler, 1966a). In a recent study by means of the ANC assay described above, Lichtenstein et al. (1967a) have confirmed many of the earlier reports based on skin test studies, as to the IgG nature of the blocking antibody.

2. Assay of Leukocyte-Sensitizing Reagins

The in vitro assay of reaginic antibody depends upon the apparently unique ability of this protein to sensitize specific target cells for histamine release. Methods dependent upon other molecular properties have proven unsuitable because of the very low concentration of these antibodies in serum. In addition, the methods which have been available, other than skin tests, do not reflect the biological activity of the antibody. Recent isolation of a reaginic antibody (IgE) and the demonstration that it possesses distinct H-chain determinants, suggest than an immunochemical method of assay may soon be available (Ishizaka et al., 1966).

The leukocyte-sensitizing ability of reaginic sera has been applied to the development of an in vitro assay (Levy and Osler, 1966). Measurement of serum reaginic activity by passive leukocyte sensitization (PLS) requires that the amount of histamine released should be a major, if not sole, function of the antibody concentration in the sensitizing mixtures. This requirement is met by performing the reaction as described in Section III above under standardized experimental conditions with respect to variables such as reaction volumes, temperature, pH, ionic composition, and antigen concentrations. Sensitization mixtures are incubated for 2 hours to provide a sufficient reaction period for weaker sera and still avoid the inhibition which has been observed with highly potent sera. After sensitization and washing, 1 ml. aliquots of the cells from each mixture are distributed into several reaction tubes, containing 1 ml. volumes of antigen E (final concentrations 1×10^{-2} and 3×10^{-2} μg./ml.) each in duplicate, or buffer.

The reagin titer (PLS_{50}) is defined as the amount of serum necessary to sensitize 10^7 leukocytes so that they will release 50% of their available histamine in the reaction with optimal amounts of antigen. Only those leukocytes capable of responding with more than 50% histamine release are suitable for these titrations. Selected dose–response curves are illustrated in Fig. 25. Normal serum controls are not routinely done, since cells sensitized with such sera release no more histamine than do unsensitized cells (Levy and Osler, 1967b). Furthermore, since the serum dose–response curve is unaltered by dilution of reaginic serum in normal serum or in buffer, the latter is routinely used as diluent (Levy and Osler, 1967b). Titrations of multiple sera from one individual are performed with leukocytes from a single nonallergic donor. Day-to-day variations, due to changes in the leukocytes of a single donor, are usually small. Thus, the titer of a reference serum in nine experiments carried out over a period of 3 months with one donor's cells varied from 23 to

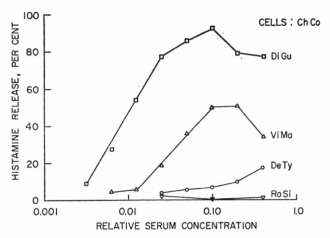

Fig. 25.   Passive leukocyte sensitization: serum dose–response relationship. Ragweed-sensitive donors, DiGu, ViMa, and DeTy; nonallergic donor, RoSi; cells, ChCo; sensitization, 180 minutes at 37°C.

34 with a mean of 28 ± 4. However, the titer of this serum varied from 12 to 159 (mean = 53 ± 35) in 19 assays performed over a 3-year period with leukocytes from six different donors. For this reason, a reference serum is assayed in each experiment to permit estimation of the relative activity of the sera under study. The ratio of PLS activity in sera assayed with leukocytes furnished by several donors is relatively constant. For example, the difference in titers of two sera tested on six sets of leukocytes ranged from 4.7 to 10.3 with a mean of 7.2 ± 1.3, an experimental error of 18% (Levy and Osler, 1967b).

When sera were assayed by the *in vitro* PLS method and in Prausnitz-Küstner tests, the activity in the skin test was almost 5 times greater than that observed with the leukocytes (Levy and Osler, 1967b).

3. Antigen Assay for Allergenic Properties

The utility of the *in vitro* reaction system has also been demonstrated in the evaluation of the allergenic activities of different ragweed pollen extracts (Lichtenstein *et al.*, 1966b). Some illustrative experiments are detailed in Figs. 26 and 27. The experiment described in Fig. 26 compared the histamine-releasing potency of two ragweed pollen preparations, a crude fraction, A, and the more highly purified Fraction IV$_2$ (King *et al.*, 1964). Two sources of leukocytes were used and they yielded the dose–response curves shown in Fig. 26. With the leukocytes of donor BWi about 10 times more of Fraction A than of Fraction IV was required to attain equivalent responses. With the cells of the second donor (JTo), Fractions A and IV$_2$ were of identical potencies. It may be reasoned that if Fraction A contained several allergens, all of which were present in Fraction IV$_2$, assays with different leukocytes should yield a constant activity ratio of A:IV$_2$. This was obviously not the case for the data (cf. Fig. 26). Variable ratios were also found for cells from six other donors. It must be concluded, therefore, that leukocytes from different donors vary in their content and specificity of those immunoglobulins that are capable of releasing histamine with ragweed pollen antigens. This interpretation is in good agreement with the varying patterns of skin reactions observed in individuals tested with the same panel of antigens (King and Norman, 1962; King *et al.*, 1964).

Further studies of the suitability of human leukocytes for *in vitro* assays of allergenic activity are described in Fig. 27 in which the bio-

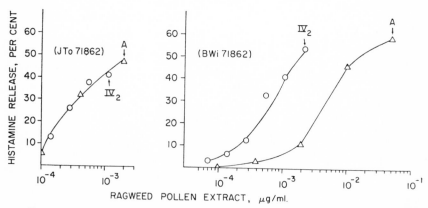

Fig. 26. Histamine release by ragweed pollen extracts A and IV$_2$. Comparative responses of leukocytes from two ragweed-sensitive donors.

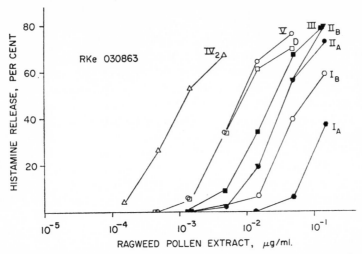

FIG. 27.   Dose–response relationships of ragweed extract Fraction D and its sub-fractions, I–V. Histamine available, 0.37 μg.; 1.6 × 10⁷ wbc.

logical activities of eight ragweed pollen fractions are compared. The range of variation is more than 100-fold. Data bearing on the sensitivity and reproducibility of these assays have been given previously (Lichtenstein et al., 1966b). In brief, the leukocyte reaction system appears to be more sensitive, more reproducible, and simpler than current methods based on skin tests of human volunteers.

B.   IMMUNOLOGICAL ASPECTS OF RAGWEED HAY FEVER

1.   Correlation of Leukocyte Responsiveness to the Severity of Hay Fever Disease

Early in the course of this investigation it was noted that cell suspensions from different hay fever patients varied markedly with respect to their reactivity with antigen E. It was suspected that this broad range of reactivity (see Fig. 11) might be related to the severity of the disease in the particular donor. An opportunity to study this relationship was afforded in a combined clinical and laboratory study (Lichtenstein et al., 1966a). The data in Fig. 28 depict a significant correlation between the severity of symptoms recorded by twenty-four patients or their physicians and the responsiveness of their leukocytes to antigen E, during the ragweed pollen season in Baltimore, 1964.

On this basis, the responsiveness of an individual's leukocytes to ragweed antigen E provides an objective and predictive measure of the clinical severity of ragweed hay fever. Cell sensitivity to antigen E is a relatively stable individual characteristic. In studies of forty patients over a 2-year period, specific therapy, or lack of it, resulted in but minor changes in the G_{50} values for most individuals. These findings do not preclude the possibility that more intensive therapy may lead to more profound changes in the level of cell responsiveness to antigen as judged by the release of histamine. VanArsdel (1965) and Pruzansky and Patterson (1967) have reported changes in cell reactivity, but the basis for these variations has not been elucidated.

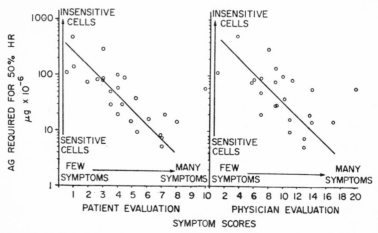

Fig. 28. The relation between *in vitro* leukocyte sensitivity as judged by duplicate titrations performed in March and April, 1964 and the symptoms these individuals suffered during the ragweed season of the same year. The patients evaluated their symptoms twice daily; the physician's evaluation is based on three visits during the ragweed season. Spearman rank correlation coefficient for patient evaluation = 0.765, $p < 0.01$; Spearman rank correlation coefficient for physician evaluation = 0.592, $p < 0.01$. (AG = antigen; HR = histamine release.)

2. Influence of Specific Therapy on the ANC in the Sera of Ragweed-Sensitive Patients

The levels of blocking activity in the sera of ragweed-sensitive individuals have been estimated during several seasons of the year, with sera from patients undergoing specific immunization with antigen E or crude ragweed extract.

The first observation of interest was that the sera of all ragweed-sensitive patients possessed demonstrable levels of ANC, i.e., blocking activity. Thus, the mean G_{50} (AHS/NHS) value for 80 sera obtained from thirty-two untreated donors was 2.9. This signifies that in the presence of these sera 3 times as much antigen was required to attain a 50% response, as compared to nonallergic individuals.

As was anticipated from earlier studies based on human skin tests (reviewed in Connell and Sherman, 1963), immunization with antigen E led to a dramatic but relatively short-lived increase in ANC. The immune response in a typical individual, immunized preseasonally over a 3-year period, is depicted in Figs. 29 and 30 (Lichtenstein *et al.*, 1968). The former shows the experimental points and a graphic demonstration of the method used in calculating the G_{50} (AHS/NHS) values (cf. also Fig. 23). With increasing immunization the antigen dose–response

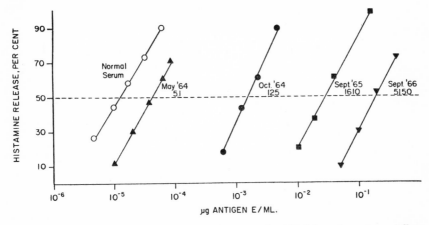

FIG. 29. Estimation of antigen E requirements for 50% histamine release following incubation of the antigen with serum specimens taken at the indicated dates.

curve, as measured with the patient's serum, shows a continual shift to the right, i.e., more and more antigen must be added to attain a 50% response. The data in Fig. 30 describe the rise and fall of ANC levels with repeated cycles of immunization with antigen E. Of interest are the time relationships in the decrease of blocking antibody activity following cessation of parenteral immunization.

During the initial year of therapy and controlled symptom evaluation, the patients receiving specific therapy showed about the same median symptom scores as did the placebo group. The following year more antigen was administered, the median antibody level rose further (about 200-fold) and with it, a significant amelioration of symptoms was reported by the group receiving antigen E (Lichtenstein *et al.*, 1968). The possible significance of these findings in relation to the assays of leukocyte-sensitizing antibodies carried out with the same serum specimens is discussed below.

3. Seasonal Variations in Leukocyte-Sensitizing Reagins

To gain further understanding of the immunology of ragweed hay fever, titrations of leukocyte-sensitizing activities were performed with sera drawn from patients before and after the ragweed pollination sea-

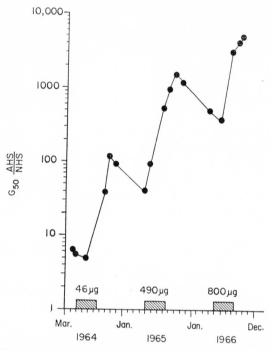

FIG. 30. Variations in antigen neutralizing capacity ("blocking" antibody) in the serum of one individual following therapeutic immunization over a 3-year period.

son. The data so obtained show that PLS_{50} titers increased during the months of September, October, and November for 21 of 25 patient-years in the Baltimore area. Serum titers tend to decline thereafter until the next pollination season. The changes in sensitizing activities observed in the serum of one patient are shown in Fig. 31. Although these data are the most dramatic that were obtained, the sera of most untreated patients showed these cyclical changes. The data in Fig. 32 illustrate the range of increases in PLS_{50} titers. The arithmetic titer ratios of postseason to preseason sera for the placebo treated patients ranged from 0.9 to 9.6,

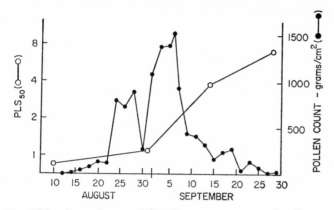

FIG. 31. PLS$_{50}$ titers of serum EdSo during 1966 ragweed pollen season. *Exp. 101766.* Cells, JoRu; sensitization, 120 minutes at 37°C.

mean-3.1. The annual contact with pollen via the upper respiratory tract seems to create a heightened sensitivity to the culpable antigen— an inference which is entirely consistent with the findings of Samter and Becker (1947), Salvaggio *et al.* (1964), Tomasi *et al.* (1965), Settipane *et al.* (1965), and Rossen *et al.* (1966) relative to the synthesis of sensitizing immunoglobulins in glandular secretions.

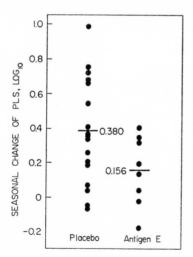

FIG. 32. Range of increase in PLS$_{50}$ titers. Ordinate: Log$_{10}$ postseason PLS titer– log$_{10}$ preseason PLS titer. Placebo: twelve patients in 17 pollen seasons of 1964–1966. Antigen E: eight patients in pollen season of 1966, $0.05 > p > 0.025$ for difference between the means, student's t test.

4. Effect of Therapeutic Immunization on the Seasonal Variations in Leukocyte-Sensitizing Reagins

The sera of ragweed-sensitive patients undergoing preseasonal therapy were also studied for changes in PLS_{50} titers. Sera were available from eight patients undergoing preseasonal, therapeutic injections with antigen E. The arithmetic titer ratios of postseasonal to preseasonal PLS_{50} titers in this treated group varied from 0.7 to 2.5, mean-1.6. These values are significantly lower than those obtained for the untreated group, $p < 0.05$, for the mean ratios. Similar differences emerged from a comparison of the mean values for the PLS_{50} assays of the treated and untreated patient groups. The mean postseason titer for the sera from the untreated patients was 34 ± 28 with a range of 1.7 to 89. For the treated patients, the corresponding values were, mean $= 10 \pm 10$ with a range of 1.3 to 27. The level of significance for this difference between the means is at $p = 0.05$. These findings, based upon the *in vitro* studies are completely in accord with the observations drawn from Prausnitz-Küstner assays by Connell and Sherman (1964a,b). Similar and previous studies by several groups of investigators indicated that skin-sensitizing antibody titers rose during the early stages of specific desensitization therapy and then fell (reviewed in Connell and Sherman, 1964a).

5. Clinical Implications

The findings accumulated during the course of this investigation may be briefly restated as follows. The severity of clinical disease in ragweed hay fever may be attributed in the main to the level of cell responsiveness to antigen, as judged by the release of histamine. This attribute reflects the quantity and activity of cell-bound reaginic immunoglobulins present in the serum and other body fluids. Upon introduction of the reactive antigen, variable portions of its allergenic activity are deflected from the target-cell sites by the circulating immunoglobulins, particularly those of the blocking type. These have been found in the sera of all ragweed hay fever patients, irrespective of therapeutic immunization. Thus, all hay fever patients synthesize at least two classes of immunoglobulins reactive with the same antigen. Those which are relatively heat-labile and show high affinity for tissue cells are the reagins, or in terms of the present studies, leukocyte-sensitizing antibodies. These are the immunoglobulins which are etiologically related to ragweed pollinosis. In contrast, the heat-stable, antigen-neutralizing or blocking antibodies are not considered to possess significant tissue affinity.

During the natural course of the disease, the levels of reaginic immunoglobulins vary with the season of the year, reaching peak activity shortly after air pollen levels fall, i.e., late September to November. The greater efficacy of transmucosal immunization in atopic individuals is indicated by the studies of Farr and Barrick (1963), Salvaggio et al. (1964), and Tomasi et al. (1965). Specific parenteral immunization alters these cyclical events. The most readily apparent effect of desensitization procedures is the elevation of the serum antigen-neutralizing capacity. From a clinical point of view, this consequence may only be palliative. The finding that the interaction of pollen antigens with these immunoglobulins follows the percentage law perhaps explains why complete symptomatic relief may occur only at extremely high antibody levels. The activity of the cell-bound sensitizing reagins seems more pertinently related to the parameter of symptom severity. Unexpectedly, it has been found that parenteral immunization in some way reduces the annual anamnestic response to airborne pollen antigens. These considerations may provide a rational basis for the difficulties encountered by previous investigators to correlate the occurrence of symptomatic cure with increased blocking antibody levels (Connell and Sherman, 1964a,b).

In the area of future studies, several goals seem attractive. From a more general, immunopathological viewpoint, elucidating the mechanism of reagin–tissue cell interaction seems the most important. The availability of purified antibody should greatly facilitate the resolution of this phenomenon. With respect to human pollinosis, the availability of monovalent, nonantigenic determinants derived from pollen antigens, which are suitable for use as a nasal spray, might provide a simple and safe means of symptomatic relief. With respect to other diseases to which immediate allergic phenomena contribute materially, the application of the human leukocyte for in vitro studies may also prove rewarding.

ACKNOWLEDGMENTS

Financial support for this work was available from the following agencies: National Science Foundation, Grant No. GB-1120; The American Cancer Society, Inc., Grant No. T-257C; The Office of The Surgeon General, Department of the Army, under the auspices of the Commission on Immunization of the Armed Forces Epidemiological Board, Contract No. DA-49-193-MD-2468; The National Institute of Allergy and Infectious Diseases, National Institutes of Health, Grant No. AI-03151-08.

REFERENCES

Andrewes, C. H., and Elford, W. J. (1933). *Brit. J. Exptl. Pathol.* 14, 367.

Arbesman, C. E., Girard, P., and Rose, N. E. (1964). *J. Allergy* 35, 535.

Archer, G. T. (1960). *Australian J. Exptl. Biol. Med.* 38, 147.

Audia, M., and Noah, J. W. (1961). *J. Allergy* 32, 223.

Austen, K. F., and Becker, E. L. (1966). *J. Exptl. Med.* 124, 397.

Austen, K. F., and Brocklehurst, W. E. (1961). *J. Exptl. Med.* 114, 29.

Austen, K. F., and Humphrey, J. H. (1962). *In* "Mechanism of Cell and Tissue Damage Produced by Immune Reactions," 2nd Intern. Symp. Immunopathol. (P. Grabar and P. Miescher, eds.), pp. 93–106. Benno Schwabe, Basel.

Austen, K. F., and Humphrey, J. H. (1963). *Advan. Immunol.* 3, 3.

Barbaro, J. F. (1961). *J. Immunol.* 86, 369, 377.

Becker, E. L., and Austen, K. F. (1966). *J. Exptl. Med.* 124, 379.

Berken, A., and Benacerraf, B. (1966). *J. Exptl. Med.* 123, 119.

Bier, O. G., Siqueira, M., and Osler, A. G. (1955). *Intern. Arch. Allergy Appl. Immunol.* 7, 1.

Binaghi, R. A., Halpern, B. N., and Liacopoulos, P. (1962). *In* "Mechanism of Cell and Tissue Damage Produced by Immune Reactions," 2nd Intern. Symp. Immunopathol. (P. Grabar and P. Miescher, eds.), pp. 123–135. Benno Schwabe, Basel.

Bloch, K. J. (1967). *Progr. Allergy* 10, 84.

Bloch, K. J., Kourilsky, F. M., Ovary, Z., and Benacerraf, B. (1963). *J. Exptl. Med.* 117, 965.

Boyden, S. V., and Sorkin, E. (1960). *Immunology* 3, 272.

Brocklehurst, W. E., Humphrey, J. H., and Perry, W. L. M. (1955). *J. Physiol. (London)* 129, 205.

Brocklehurst, W. E., Humphrey, J. H., and Perry, W. L. M. (1961). *Immunology* 4, 67.

Chakravarty, N. (1960). *Acta Physiol. Scand.* 48, 146.

Citron, J. (1908). *Berlin Klin. Wochschr.* 45, 518.

Coca, A. F., and Grove, E. F. (1925). *J. Immunol.* 10, 445.

Cohn, Z. A., and Parks, E. (1967). *J. Exptl. Med.* 125, 213.

Colquhoun, D., and Brocklehurst, W. E. (1965). *Immunology* 9, 591.

Connell, J. T., and Sherman, W. B. (1963). *J. Immunol.* 91, 187.

Connell, J. T., and Sherman, W. B. (1964a). *J. Allergy* 35, 18.

Connell, J. T., and Sherman, W. B. (1964b). *J. Allergy* 35, 169.

Cooke, R. A., Barnard, J. H., Hebald, S., and Stull, A. J. (1935). *J. Exptl. Med.* 62, 733.

Cooke, R. A., Loveless, M., and Stull, A. (1937). *J. Exptl. Med.* 66, 689.

Copenhaver, J. H., Nagler, M. E., and Goth, A. (1953). *J. Pharmacol. Exptl. Therap.* 109, 401.

Dale, H. H. (1913). *J. Pharmacol. Exptl. Therap.* 4, 167.

de St. Groth, F. (1962). *Advan. Virus Res.* 9, 1.

Dias da Silva, W., and Lepow, I. H. (1965). *J. Immunol.* 95, 1080.

Farr, R. S., and Barrick, R. H. (1963). *Proc. 19th Ann. Meeting, Am. Acad. Allergy, Montreal* (abstr.).

Fawcett, D. W. (1954). *J. Exptl. Med.* 100, 217.

Frick, O. L., Liacopoulos, P., and Raffel, S. (1965). *J. Immunol.* **94**, 890.

Friedberger, E. (1909). *Z. Immunitaetsforsch.* **2**, 208.

Gocke, D. J., and Osler, A. G. (1965). *J. Immunol.* **94**, 236, 247.

Graham, H. T., Lowry, O. H., Wheelwright, F., Lenz, M. A., and Parish, H. H., Jr. (1955). *Blood* **10**, 467.

Green, H., and Goldberg, B. (1960). *Ann. N. Y. Acad. Sci.* **87**, 352.

Halpern, B. N., Liacopoulos, P., Liacopoulos-Briot, M., and Binaghi, R. (1958). *Compt. Rend. Soc. Biol.* **247**, 1798.

Humphrey, J. H., and Jaques, R. (1955). *J. Physiol. (London)* **128**, 9.

Humphrey, J. H., Austen, K. F., and Rapp, H. J. (1963). *Immunology* **6**, 226.

Ishizaka, K., and Ishizaka, T. (1966). *J. Allergy* **38**, 108.

Ishizaka, K., Ishizaka, T., and Sugahara, T. (1961). *J. Immunol.* **87**, 548.

Ishizaka, K., Ishizaka, T., and Lee, E. H. (1966). *J. Allergy* **37**, 336.

Jensen, J. (1967). *Science* **155**, 1122.

Katz, G., and Cohen, S. (1941). *J. Am. Med. Assoc.* **117**, 1782.

Keller, R. (1966). "Tissue Mast Cells in Immune Reactions." Elsevier, New York.

King, T. P., and Norman, P. S. (1962). *Biochemistry* **1**, 709.

King, T. P., Norman, P. S., and Connell, J. T. (1964). *Biochemistry* **3**, 458.

Kremzner, L. T., and Wilson, I. B. (1961). *Biochim. Biophys. Acta* **50**, 364.

Lapin, J. H., Horonock, A., and Lapin, R. H. (1958). *Blood* **13**, 1001.

Layton, L. (1965). *J. Allergy* **36**, 523.

Layton, L. L., Yamanaka, E., Greene, F. C., and Perlman, F. (1963). *Intern. Arch. Allergy Appl. Immunol.* **23**, 87.

Leddy, J. P., Freeman, G. L., Luz, A., and Todd, R. H. (1962). *Proc. Soc. Exptl. Biol. Med.* **111**, 7.

Levy, D. A., and Osler, A. G. (1966). *J. Immunol.* **97**, 203.

Levy, D. A., and Osler, A. G. (1967a). *J. Immunol.*, in press.

Levy, D. A., and Osler, A. G. (1967b). *J. Immunol.*, in press.

Levy, D. A., and Osler, A. G. (1967c). Unpublished observations.

Lichtenstein, L. M., and Osler, A. G. (1964). *J. Exptl. Med.* **120**, 507.

Lichtenstein, L. M., and Osler, A. G. (1966a). *J. Immunol.* **96**, 159.

Lichtenstein, L. M., and Osler, A. G. (1966b). *J. Immunol.* **96**, 169.

Lichtenstein, L. M., and Osler, A. G. (1966c). *Proc. Soc. Exptl. Biol. Med.* **121**, 808.

Lichtenstein, L. M., Norman, P. S., Winkenwerder, W. L., and Osler, A. G. (1966a). *J. Clin. Invest.* **45**, 1126.

Lichtenstein, L. M., King, T. P., and Osler, A. G. (1966b). *J. Allergy* **38**, 174.

Lichtenstein, L. M., Holtzman, N. A., and Margolis, S. (1967). *Federation Proc.* **26**, 306 (abstr.).

Lichtenstein, L. M., Norman, P. S., and Winkenwerder, W. L. (1968). *Am. J. Med.*, in press.

Loveless, M. H. (1940). *J. Immunol.* **38**, 25.

Matthews, K. P., Farrar, M. O., and Burleigh, E. B. (1962). *Proc. 18th Ann. Meeting, Am. Acad. Allergy, Denver, Colo.* (abstr.).

May, C. D., Cheng, J., and Lymar, M. (1967). *J. Allergy* **39**, 123 (abstr.).

Mayer, M. M. (1965). *Ciba Found. Symp. Complement* pp. 1–32.

Middleton, E. (1960). *Proc. Soc. Exptl. Biol. Med.* **104**, 245.

Middleton, E., and Sherman, W. B. (1960). *J. Allergy* **31**, 441.

Middleton, E., Sherman, W. B., Fleming, W., and VanArsdel, P. P. (1960). *J. Allergy* **31**, 448.

Mills, S. E., and Levine, L. (1959). *Immunology* **2**, 368.
Mongar, J. L., and Schild, H. O. (1956). *J. Physiol.* (*London*) **131**, 207.
Mongar, J. L., and Schild, H. O. (1957). *J. Physiol.* (*London*) **135**, 320.
Mongar, J. L., and Schild, H. O. (1958). *J. Physiol.* (*London*) **140**, 272.
Mongar, J. L., and Schild, H. O. (1960). *J. Physiol.* (*London*) **150**, 546.
Mongar, J. L., and Schild, H. O. (1962). *Physiol. Rev.* **42**, 226.
Mota, I. (1963). *Ann. N. Y. Acad. Sci.* **103**, 264.
Müller-Eberhard, H. (1966). *Arch. Pathol.* **82**, 205.
Neu, H. C., Randall, H. G., and Osler, A. G. (1961). *Immunology* **4**, 401.
Nielsen, C. B., Terres, G., and Feigen, G. A. (1959). *Science* **130**, 41.
Noah, J. W., and Brand, A. (1954). *J. Allergy* **25**, 210.
Noah, J. W., and Brand, A. (1955). *J. Allergy* **26**, 385.
Noah, J. W., and Brand, A. (1963). *J. Allergy* **34**, 203.
Osler, A. G., Hawrisiak, M. M., Ovary, Z., Siqueira, M., and Bier, O. G. (1957). *J. Exptl. Med.* **106**, 811.
Osler, A. G., Randall, H. G., Hill, B. M., and Ovary, Z. (1959a). *J. Exptl. Med.* **110**, 311.
Osler, A. G., Randall, H. G., Hill, B. M., and Ovary, Z. (1959b). *In* "Mechanisms of Hypersensitivity," Intern. Symp. (J. H. Shaffer, G. A. LoGrippo, and M. W. Chase, eds.), pp. 281–304, Little, Brown, Boston, Massachusetts.
Osler, A. G., Lichtenstein, L. M., and Levy, D. A. (1964). *Arch. Exptl. Pathol. Pharmakol.* **250**, 111.
Ovary, Z., and Karush, F. (1961). *J. Immunol.* **86**, 146.
Ovary, Z., Benacerraf, B., and Bloch, K. J. (1963). *J. Exptl. Med.* **117**, 951.
Prausnitz, C., and Küstner, H. (1921). *Centr. Bakteriol. Parasitenk. Infekt.* **86**, 160.
Pruzansky, J. J., and Patterson, R. (1966). *J. Allergy* **38**, 315.
Pruzansky, J. J., and Patterson, R. (1967). *J. Allergy* **39**, 44.
Rose, N. E., Kent, J. H., Reisman, R. E., and Arbesman, C. E. (1964). *J. Allergy* **35**, 525.
Rossen, R. D., Alford, R. H., Butler, W. T., and Kasel, J. A. (1966). *J. Clin. Invest.* **45**, 768.
Salvaggio, J. E., Cavanaugh, J. J. A., Lowell, F. C., and Leskowitz, S. (1964). *J. Allergy* **35**, 62.
Samter, M., and Becker, E. L. (1947). *Proc. Soc. Exptl. Biol. Med.* **65**, 140.
Schild, H. O. (1939). *J. Physiol.* (*London*) **95**, 393.
Secchi, A. G., Siraganian, R. P., and Osler, A. G. (1966). Unpublished observations.
Secchi, A. G., Siraganian, R. P., Dorsch, C., and Osler, A. G. (1967). Unpublished observations.
Settipane, G. A., Connell, J. T., and Sherman, W. B. (1965). *J. Allergy* **36**, 92.
Shore, P. A., Burkhalter, A., and Cohn, V. H., Jr. (1959). *J. Pharmacol. Exptl. Therap.* **127**, 182.
Siraganian, R. P., Secchi, A. G., and Osler, A. G. (1967). *Federation Proc.* **26**, 309 (abstr.).
Smith, D. E. (1958a). *Science* **128**, 207.
Smith, D. E. (1958b). *Am. J. Physiol.* **193**, 573.
Smith, D. E. (1963). *Ann. N. Y. Acad. Sci.* **103**, 40.
Stanworth, D. R. (1963). *Advan. Immunol.* **3**, 181.
Stechschulte, D. J., Austen, K. F., and Bloch, K. J. (1967). *J. Exptl. Med.* **125**, 127.

Stroud, R. M., Mayer, M. M., Miller, J. A., and McKenzie, A. T. (1966). *Immunochemistry* **3**, 163.

Tomasi, T. B., Jr., Tan, E. M., Solomon, A., and Predergast, R. A. (1965). *J. Exptl. Med.* **121**, 101.

Uhr, J. W. (1965). *Proc. Natl. Acad. Sci. U.S.* **54**, 1599.

VanArsdel, P. P., Jr. (1965). *Federation Proc.* **24**, 632 (abstr.).

VanArsdel, P. P., Jr., and Middleton, E. (1961). *J. Allergy* **32**, 348.

VanArsdel, P. P., Jr., and Sells, C. J. (1963). *Science* **141**, 1190.

VanArsdel, P. P., Jr., Middleton, E., Sherman, W. B., and Buchwald, H. (1958). *J. Allergy* **29**, 429.

Ward, P. A., Cochrane, C. G., and Müller-Eberhard, H. J. (1965). *J. Exptl. Med.* **122**, 327.

Wilkins, D. J., Ottewill, R. H., and Bangham, A. D. (1962). *J. Theoret. Biol.* **2**, 165, 176.

Yagi, Y., Maier, P., Pressman, D., Arbesman, C. E., and Reisman, R. E. (1963). *J. Immunol.* **91**, 83.

AUTHOR INDEX

Numbers in italics refer to pages on which the references are listed.

SUBJECT INDEX

A

Anaphylatoxin, formation of, 55–58
Antibody,
  antigen metabolism and, 95–100
  immune cytolysis and, 25–28
  suppression of antibody-synthesizing
    cells by, 100–110
Antibody formation,
  increase, antigen-antibody complex
    and, 94–95
  suppression by antibody,
    antibody fragments, 93–94
    different classes of antibody, 91–93
    primary response, 85–89
    priming, 89–90
    secondary response, 90
Antibody-synthesizing cells, suppression
    by antibody, 100–110
Antigen,
  catabolism, immune response and,
    97–100
  competition of, 148
  fetal, maternal paralysis by, 148
  histamine release from leukocytes and,
    198–204
  metabolism, antibody and, 95–100
  overloading, immunological paralysis
    and, 143–146
  removal, antibody and, 95–97
Antigen-antibody complex, antibody for-
    mation and, 94–95
Autoimmunity,
  experimental, suppression of, 110–111

B

Bacteria, complement and, 54–55

C

Castration, complement deficiency and,
  69–70
Clots,
  whole-blood, lysis *in vitro*, 64–65

Cobra factor, complement deficiency and,
  70–71
Complement,
  anaphylatoxin formation and, 55–58
  bacteria and, 54–55
  components,
    highly purified, 4–11
    partially purified, 11–14
    preparation of, 14–22
    properties of, 4–14
    quantitation of, 48–52
  effects, ultrastructure and, 23–24
  enzymatic activities and, 46–48
  experimentally induced deficiencies,
    cobra factor, 70–71
    sex hormones and castration, 69–70
  highly purified components,
    preparation of, 14–18
    properties of, 4–11
  immune adherence and opsonization,
    61–62
  immune cytolytic effect,
    antibody and, 25–28
    attachment of C'3 and, 38–40
    components C'2–C'4 and, 31–38
    first component and, 28–31
    schematic description of, 24–25
    stable intermediate complex and,
      40–43
    terminal steps of, 43–46
    ultrastructure and, 22–24
  immunoconglutination and conglutina-
    tion, 62–64
  inherited deficiencies,
    guinea pig, 65
    man, 68–69
    mouse, 65–66
    rabbit, 67–68
  leukocyte chemotactic factors and, 58–
    59
  nomenclature and, 3–4
  noncytolytic reactions, mechanism of,
    55–65